The Einsteinian Era

ii

The Einsteinian Era

Frank W. K. Firk

Professor Emeritus of Physics

Yale University

iv

ISBN – 13: 9781499351989

ISBN – 10:1499351984

CONTENTS

PREFACE

This book has its origin in a one-year course that I taught at Yale throughout the decade of the 1970's. The course was for non-science majors who were interested in learning about the major branches of Physics. In the first semester, emphasis was placed on Newtonian and Einsteinian Relativity. The recent popularity of the Einstein exhibit at the American Museum of Natural History in New York City, prompted me to look again at my fading lecture notes. I found that they contained material that might be of interest to today's readers. I have therefore reproduced them with some additions, mostly of a graphical nature. I recall the books that were most influential in my approach to the subject at that time; they were Robert Adair's *Concepts of Physics*, and Casper and Noer's *Revolutions in Physics*. These books were written with the non-scientist in mind, and they showed what could be achieved in this important area of teaching and learning; I am greatly indebted to these authors.

Guilford, Connecticut, 2003

1. INTRODUCTION

This brief book is for the inquisitive reader who wishes to gain an understanding of the immortal work of Einstein, the greatest scientist since Newton. Special Relativity deals with measurements of space, time and motion in inertial frames of reference (see chapter 4). An introduction to Einstein's Theory of General Relativity, a theory of space, time and motion in the presence of gravity, is given at a popular level. A more formal account of Special Relativity that requires a higher level of understanding of Mathematics is given in an Appendix.

Historians in the future will, no doubt, choose a phrase that best characterizes the 20th-century. I believe that a strong case will be made for the phrase "The Einsteinian Era"; no other person in the 20th-century advanced our understanding of the physical universe in such a dramatic way. He introduced many original concepts, each one of a profound nature. His discovery of the universal equivalence of energy and mass has had, and continues to have, far-reaching consequences not only in Science and Technology but also in fields as diverse as World Politics, Economics, and Philosophy. The topics covered include:

1) Understanding the physical universe

2) Describing everyday motion:

   relative motion,

   Newton's Principle of Relativity,

   problems with light,

3) Einstein's Theory of Special Relativity:

simultaneity and synchronizing clocks,

length contraction and time dilation,

examples of Einstein's world,

4) Newtonian and Einsteinian mass

5) Equivalence of energy and mass, $E = mc^2$

6) Principle of Equivalence

7) Einsteinian gravity:

gravity and the bending of light,

gravity and the flow of time,

red shifts, blue shifts, and black holes.

2. UNDERSTANDING THE PHYSICAL UNIVERSE

We would be justified in thinking that any attempts to derive a small set of fundamental

laws of Nature from a limited sample of all possible processes in the physical universe, would

lead to a large set of unrelated facts. Remarkably, however, very few fundamental laws of

Nature have been found to be necessary to account for all observations of basic physical

phenomena. These phenomena range in scale from the motions of minute subatomic systems

to the motions of the galaxies. The methods used, over the past five hundred years, to find the

set of fundamental laws of Nature are clearly important; a random approach to the problem

would have been of no use whatsoever. In the first place, it is necessary for the scientist to have

a conviction that Nature can be understood in terms of a small set of fundamental laws, and that these laws should provide a quantitative account of all basic physical processes. It is axiomatic that the laws hold throughout the universe. In this respect, the methods of Physics belong to Philosophy. (In earlier times, Physics was referred to by the appropriate title, "Natural Philosophy").

2.1 Reality and Pure Thought

In one of his writings entitled "On the Method of Theoretical Physics", Einstein stated: "If, then, experience is the alpha and the omega of all our knowledge of reality, what then is the function of pure reason in science?" He continued, "Newton, the first creator of a comprehensive, workable system of theoretical physics, still believed that the basic concepts and laws of his system could be derived from experience." Einstein then wrote "But the tremendous practical success of his (Newton's) doctrines may well have prevented him, and the physicists of the eighteenth and nineteenth centuries, from recognizing the fictitious character of the foundations of his system". It was Einstein's view that " ... the concepts and fundamental principles which underlie a theoretical system of physics are *free inventions of the human intellect*, which cannot be justified either by the nature of that intellect or in any other fashion *a priori*." He continued, "If, then, it is true that the axiomatic basis of theoretical physics cannot be extracted from experience but must be freely invented, can we ever hope to find the right way? ... Can we hope to be guided safely by experience at all when there exist theories (such as Classical (Newtonian) Mechanics) which, to a large extent, do justice to experience without getting to the root of the matter? I answer without hesitation that there is, in

4

my opinion, a right way, and that we are capable of finding it." Einstein then stated

"Experience remains, of course, the sole criterion of the physical utility of a mathematical

construction. But the creative principle resides in Mathematics. I hold it true that pure thought

can grasp reality, as the ancients dreamed." This is, of course, an extreme philosophical

position.

3. DESCRIBING EVERYDAY MOTION

3.1 Motion in a straight line (the absence of forces)

The simplest motion is that of a point, P, moving in a straight line. Let the line be

labeled the "x-axis", and let the position of P be measured from a fixed point on the line, the

origin, O. Let the motion begin (time $t = 0$) when P is at the origin ($x = 0$). At an arbitrary time,

t, P is at the distance x:

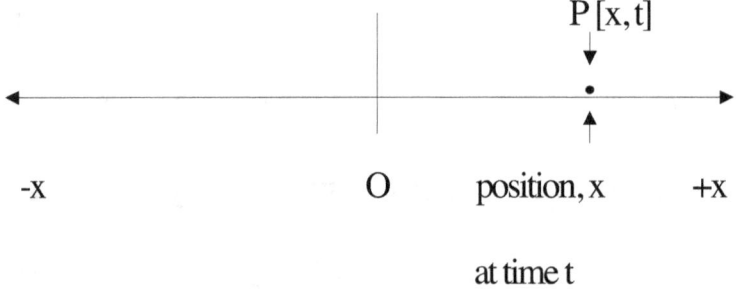

If successive positions of P are plotted, together with their corresponding times, we can

generate what is called the "world line" of P.

Let us observe a racing car moving at high speed along the straight part of a race track

(the x-axis), and let us signal the instant that it passes our position, $x = 0$, by lowering a flag:

An observer, standing at a measured distance D, from x = 0, starts his clock at the instant, t = 0, when he sees the flag lowered, and stops his clock at the instant t = T, as the car passes by. We can obtain the average speed of the car, v, during the interval T, in the standard way; it is

$$v = D/T.$$

If, for example, D = 1 mile, and T = 20 seconds (1/180 hour) then

v = 1 (mile)/(1/180) (hour) = 180 miles per hour.

This is such a standard procedure that we have no doubt concerning the validity of the result.

3.2 The relativity of everyday events

Events, the description of when and where happenings occur, are part of the physical world; they involve finite extensions in both time and space. From the point of view of a theory of motion, it is useful to consider "point-like" events that have vanishingly small extensions in time and space. They then can be represented as "points" in a spacetime geometry. We shall label events by giving the time and space coordinates: event E → E[t, x], or in three space dimensions, E[t, x, y, z], where x, y, z are the Cartesian components of the position of the event. There is nothing special about a Cartesian coordinate system, it is a mathematical construct; *any* suitable coordinate mesh with a metrical property (measured distances defined in terms of coordinates) can be used to describe the spatial locations of events. A familiar non-Cartesian system is the spherical polar coordinate system of the lines of latitude and longitude on the surface of the earth. The time t can be given by any device that is capable of producing a stable, repetitive motion such as a pendulum, or a quartz-controlled crystal oscillator or, for high precision, an atomic clock.

Suppose we have an observer, O, at rest at the origin of an x– axis, in the F - frame. O has assistants with measuring rods and clocks to record events occurring on the x-axis:

We introduce a second observer, O′, at rest at the origin of his frame of reference, F ′. O′ has his assistants with their measuring rods (to measure distances, x′) and clocks (to measure times, t′) to record events on the x′ - axis. (The F ′- clocks are identical in construction and performance to the clocks in the F-frame). Let O′ coincide with O at a common origin O = O′ (x = x′ = 0), at the synchronized time zero t = t′ = 0. At t = t′= 0, we have

Suppose that the observer O′, and his assistants with rods and clocks, move to the right with *constant speed* V along the common x, x′ - axis. At some later time t, the two sets of observers, represented by O and O′, record a common event that they write as E[t, x] and E ′[t′, x′], respectively. The relative positions of the two observers at time t is:

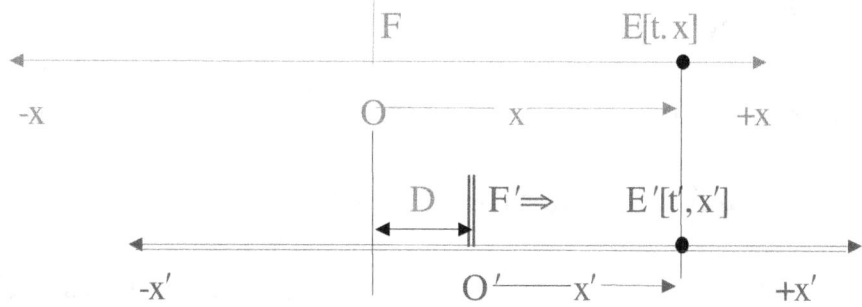

where D = Vt is the distance that O′ moves at constant speed V, in the time t.

We therefore write the relationship between the two measurements by the plausible equations (based on everyday experience):

$$t' = t \text{ (everyday identical clocks tick at the same rate)}$$

$$x' = x - D = x - Vt.$$

These are the basic equations of relative motion according to the concepts first put forward by Galileo and Newton. They are fully consistent with measurements made in our real world (the world of experience). They are not, however, *internally consistent*. In the equation that relates the measurement of distance x′ in the F′- frame to the measurements in the F - frame, we see that the space part, x′, in the F′- frame, is related to the space part, x, *and* the time part, t, in the F - frame: spacetime in one frame is not related to spacetime in the other frame! Furthermore, the time equation makes no mention of space in either frame. *We see that there is a fundamental lack of symmetry in the equations of relative motion, based on everyday experience.* The question of the "symmetry of spacetime" will lead us to Einstein's philosophy of the "free invention of the intellect".

3.3 Relative velocities

We have seen that the position of an event, E[t, x], measured by an observer O, is related to the position of the same event, E'[t', x'], measured by an observer O', moving with constant speed V along the common x, x' – axis of the two frames, by the equation

$$x' = x - Vt .$$

The speed v of a point P[t, x], moving along the x – axis, is given by the ratio of the finite distance the point moves, Δx, in a given finite time interval, Δt:

$$v = \Delta x / \Delta t .$$

We can obtain the speeds v, and v' of the same moving point, as measured in the two frames, by calculating $v = \Delta x / \Delta t$ and $v' = \Delta x' / \Delta t'$, as follows:

$$\Delta x' / \Delta t' = v' = \Delta x / \Delta t - V \Delta t / \Delta t \text{ (where we have used } \Delta t' = \Delta t \text{ because } t = t' \text{ in everyday}$$

experience).

We therefore find

$$v' = v - V,$$

the speeds differ by the relative speed of the two frames. This is consistent with experience: if a car moves along a straight road at a constant speed of v = 60 mph, relative to a stationary observer O, and an observer O' follows in a car at a constant speed of V = 40 mph relative to O, then the speed of the first car relative to the occupant of the second car is v' = 20 mph.

3.4 The Newtonian Principle of Relativity

The Newtonian *Principle of Relativity* asserts that, in the inertial frames F, F ′, the

following two situations

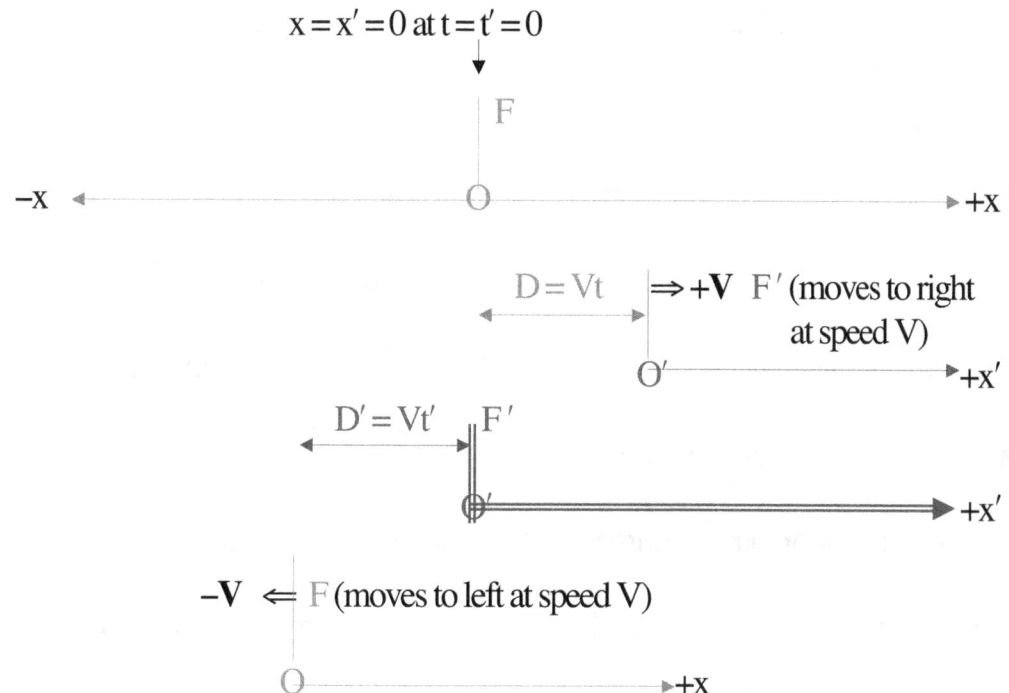

cannot be distinguished by experiments that involve mechanical systems (classical systems

that obey Newton's Laws of Motion).

The speed V has been written in **bold face** to remind us that here we are dealing with a

vector quantity that has both magnitude (the speed in mph) and a sense of direction: +**V** in the

+x - direction and −**V** in the −x - direction.

3.5 Problems with light

We are accustomed to the notion that waves propagate through a medium, required to

support the waves. For example, sound waves propagate as pressure variations in air, and water waves propagate as coupled displacements of the water molecules, perpendicular to the direction of the wave motion. In the 19th-century, Maxwell discovered that light waves are electromagnetic phenomena. This great work was based on theoretical arguments, motivated by the experimental results of Faraday and Henry. One of the most pressing questions facing scientists of the day was:

"what is waving when a beam of light propagates through empty space?"

It was proposed that the universe is filled with a medium called the *aether* with the property of supporting light waves, and having no other physical attributes. (For example, it would have no effect on the motion of celestial bodies). In the latter part of the 19th-century, Michelson and Morley carried out a famous experiment at the Case Institute in Cleveland that showed there is *no experimental evidence for the aether*. Light travels through the void, and that is that. Implicit in their work was the counter - intuitive notion that the speed of light does not depend on the speed of the source of the light.

The *Aether Theory* was popular for many years. Non - traditional theories were proposed to account for the null - result of the Michelson - Morley experiment. Fitzgerald (Trinity College, Dublin) proposed that the Michelson - Morley result could be explained, and the Aether Hypothesis retained, if the lengths of components in their apparatus were "velocity-dependent" – lengths contract in the direction of motion, and lengths remain unchanged when perpendicular to the direction of motion. He obtained the result

$$L_0 \quad = \quad [1/\sqrt{(1-(v/c)^2)}]\, L \quad \equiv \quad \gamma L$$

⇑ ⇑

(length of rod at rest) (length of rod moving at speed v)

Here, c is the constant speed of light ($2.99\ldots \times 10^8$ meters/second).

All experiments are consistent with the statement that the ratio v/c is always less than 1, and therefore γ is always greater than 1. This means that the measured length of the rod L_0, in its rest frame, is always greater than its measured length when moving.

At the end of the 19th - century, Larmor introduced yet another radical idea: a moving clock is observed to tick more slowly than an identical clock at rest. Furthermore, the relationship between the clock rates in the moving and rest frames is given by the same factor, γ, introduced previously by Fitzgerald. Specifically,

$$\Delta t \quad = \quad \gamma \Delta t_0$$

⇑ ⇑

(an interval on a moving clock) (an interval on a clock at rest)

Since the velocity-dependent term γ is greater than 1, the intervals of time Δt (moving), are greater than the intervals Δt_0 (at rest).

Fitzgerald, Larmor, and other physicists at that time considered length contraction and time dilation to be "real" effects, associated with minute physical changes in the structure of rods and clocks when in motion. It was left to the young Einstein, working as a junior Patent Officer in Bern, and thinking about space, time, and motion in his spare time, to introduce a new Theory of Relativity, uninfluenced by the current ideas. Although, in later life, Einstein acknowledged that he was aware of the Michelson - Morley result, and of the earlier work on length contraction and time dilation, he neither used, nor referred to, the earlier results in his first paper on Relativity.

4. EINSTEIN'S THEORY OF SPECIAL RELATIVITY

In 1905, Einstein published three great papers in unrelated areas of Physics. In this chapter, we shall discuss his new ideas concerning the relative motion of beams of light, and of objects that move at speeds close to that of light. His independent investigations were based upon just two postulates:

1. The generalized *Principle of Relativity*: no *mechanical or optical* experiments can be carried out that distinguish one inertial frame of reference from another. (This is a development of Newton's Principle of Relativity that is limited to *mechanical* experiments, involving speeds much less than the speed of light; it applies to experiments in the everyday world).

Inertial frames of reference are non-rotating, and move in straight lines at constant speed. They are non-accelerating.

2. The speed of light in a vacuum is a constant of Nature, and is independent of the velocity of the source of the light.

Einstein was not concerned with questions having to do with the Aether; for him, a true theory of the physical properties of the universe could not rest upon the mysterious qualities of such an *unobservable.* As we shall see, Einstein was concerned with the precise meaning of *measurements* of lengths and time intervals. In his later years, he recalled an interesting thought that he had while in school. It had to do with the meaning of time. Our lives are dominated by "psychological time"; for example, time seems to go by more quickly as we grow older. For the young Einstein, *time* in the physical world was simply the reading on a clock. He therefore imagined the following: if the schoolroom clock is reading 3PM, and I rush away from the clock at the speed of light, then the information (that travels at the speed of light) showing successive ticks on the clock, and therefore the passage of time, will never reach me, and therefore, in my frame of reference, it is forever 3PM – time stands still. He therefore concluded that the measurement of time must depend, in some way, on the relative motion of the clock and the observer; he was, by any standards, a precocious lad.

If we apply the Galilean-Newtonian expression for the relative velocities v, v′, measured in the inertial frames F, F′, (moving with relative speed V), to the measurements of flashes of light, v = c, the speed of light in F, and v′ = c′, the speed of light in F′, we expect

c′ = c – V (corresponding to v′ = v – V for everyday objects).

Modern experiments in Atomic, Nuclear, and Particle Physics are consistent with the fact that c′ = c, no matter what the value of the relative speed V happens to be.

At the end of the 19th - century, a key question that required an answer was therefore: - why does the Galilean - Newtonian equation, that correctly describes the relative motion of everyday objects, fail to describe the relative motion of beams of light? Einstein solved the problem in a unique way that involved a fundamental change in our understanding of the

nature of space and time, a change that resulted in far - reaching consequences; these

consequences are discussed in the following chapters.

4.1 The relativity of simultaneity: the synchronization of clocks.

It is important to understand the meaning of the word "observer" in Relativity. To

record the time and place of a sequence of events in a particular inertial reference frame, it is

necessary to introduce an *infinite set* of adjacent "observers", located throughout the entire

space. Each observer, at a known, fixed position in the reference frame, carries a clock to

record the time, and the characteristic property, of every event in his immediate neighborhood.

The observers are not concerned with non-local events. The clocks carried by the observers

are *synchronized*: they all read the *same time* throughout the reference frame. It is the job of the

chief observer to collect the information concerning the time, place, and characteristic feature

of the events recorded by all observers, and to construct the world line (a path in space-time),

associated with a particular characteristic feature (the type of particle, for example). "Observer"

is therefore seen to be a *collective* noun, representing the infinite set of synchronized observers

in a frame of reference.

The clocks of all observers in a reference frame are synchronized by correcting them

for the speed of light (the speed of information) as follows:

Consider a set of clocks located at x_0, x_1, x_2, x_3, along the x-axis of a reference frame.

Let x_0 be the position of the chief observer, and let a flash of light be sent from the clock at x_0

when it reads t_0 (12 noon, say). At the instant that the light signal reaches the clock at x_1, it is set to read $t_0 + (x_1/c)$, at the instant that the light signal reaches the clock at x_2, it is set to read $t_0 + (x_2/c)$, and so on for every clock along the x-axis. All clocks along the x - axis then "read the same time" – they are synchronized:

These 4 clocks read the same time "noon + x_3/c in their rest frame

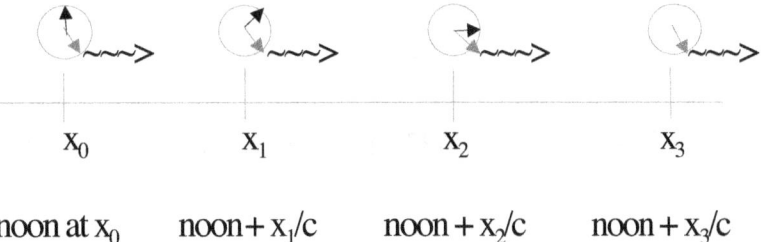

| noon at x_0 | noon + x_1/c | noon + x_2/c | noon + x_3/c |

To all other inertial observers, the clocks appear to be *unsynchronized*.

The relativity of simultaneity is clearly seen using the following method to synchronize two clocks: a flash of light is sent out from a source, M′ situated midway between identical clocks, A′ and B′, *at rest* in the frame, F′

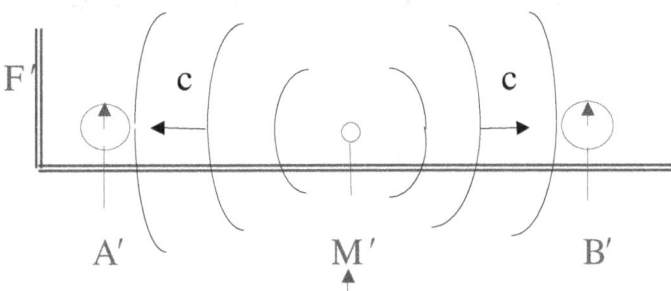

Flash of light from M′, the mid-point between A′ and B′

The two clocks are synchronized by the (simultaneous) arrival of the flash of light (traveling at c) from M′. We now consider this process from the viewpoint of observers in an inertial

16

frame, F, who observe the F ′-frame to be moving to the right with constant speed V. From

their frame, the synchronizing flash reaches A′ before it reaches B′:

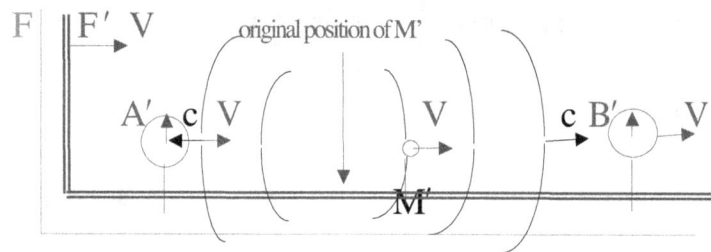

They conclude, therefore, that the A′-clock starts before the B′-clock; the clocks are no longer

synchronized. This analysis rests on the fact that the speed of light does not depend on the

speed of the source of light.

The relativity of simultaneity leads to two important non-intuitive results namely,

length contraction and *time dilation.*

4.2 Length contraction

Let a rod be at rest in the F-frame, and let its proper (rest) length be L_0.

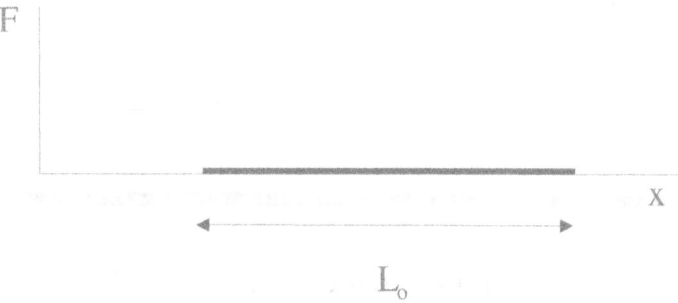

Consider an F ′- frame, moving at constant speed V in the +x direction. The set of observers, at

rest in F ′, have synchronized clocks in F ′, as shown

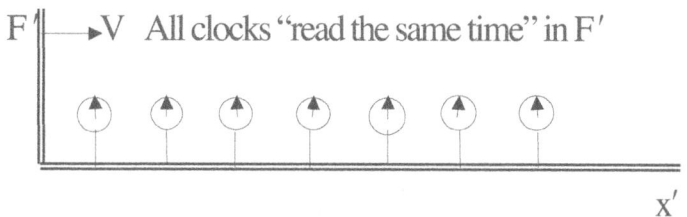

The observers in F' wish to determine the length of the rod, L', relative to the F' frame. From their perspective, the rod is moving to the left (the $-x'$ direction) with constant speed, V. We define the length of a rod, measured in any inertial frame, in terms of the positions of the two ends of the rod *measured at the same time*. If the rod is at rest, it does not matter when the two end positions are determined; this is clearly not the case when the rod is in motion. The observers in F' are distributed along the x'–axis, as shown. They are told to measure the length of the rod at 12 noon. This means that, as the rod passes by, each observer looks to see if either end of the rod is in his (immediate) vicinity. If it is, the two critical observers A' and B' (say) raise their hands. At any time later, the observers in F' measure the distance between the observers A' and B', and the chief observer states that this is the length of the rod, in their frame. This procedure can be carried out only if the clocks in F' are synchronized. We have seen, however, that the synchronized clocks in F' are not seen to be synchronized in a different frame, F, such that F' moves at speed V, relative to F. The question is: how does the length L' of the moving rod, determined by the observers in F', appear to the observers at rest in the F frame? We have seen that the clock A' starts before clock B', according to the F observers. Therefore "left end of the rod coincides with the A' clock, reading noon" occurs before "right end of rod coincides with B' clock, reading noon", according to the F observers:

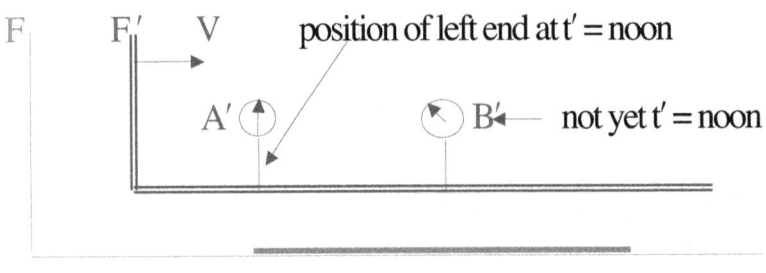

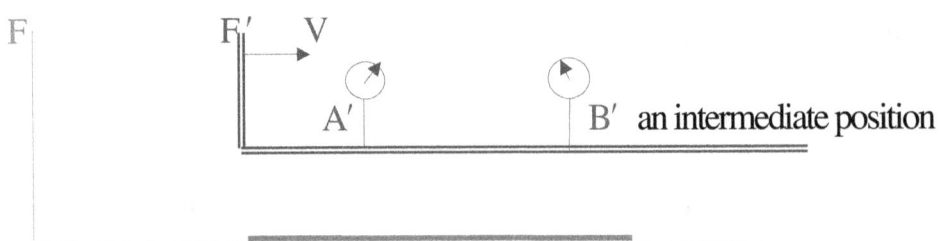

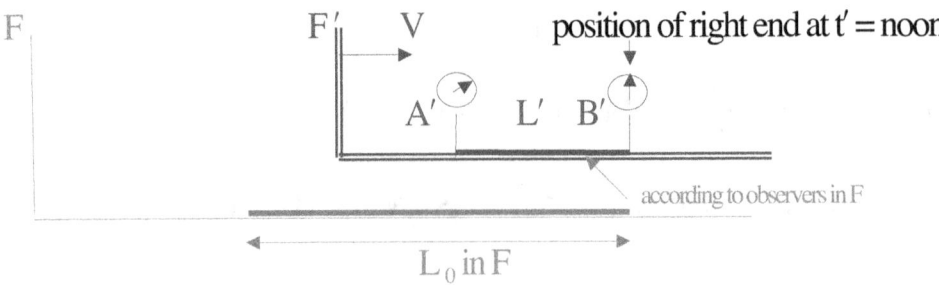

The length of the rod, L′, as determined by observers at rest in F′, *shown from the viewpoint of observers in the F frame*, is less than its proper length, L_0.

We see that the measurement of length contraction does not involve physical changes in a moving rod; it is simply a consequence of the synchronization of clocks in inertial frames, and the relativity of simultaneity.

4.3 Time dilation

The *proper time* interval between two events is the interval measured in the frame of reference in which they occur *at the same position*. Intervals that take place at different positions are said to be *improper*.

Consider a pulse of light that reflects between two plane mirrors, M_1 and M_2, separated by a distance D:

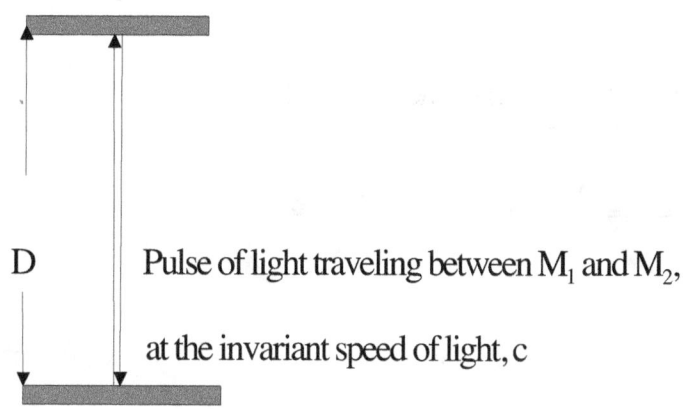

D   Pulse of light traveling between M_1 and M_2,

at the invariant speed of light, c

The time interval, Δt, for the light to make the round-trip is $\Delta t = 2D/c$.

Consider a reference frame, F, (the laboratory frame, say), and let the origin of F coincide with the location of an event E_1. A second event E_2, occurs at a different time and location in F, thus

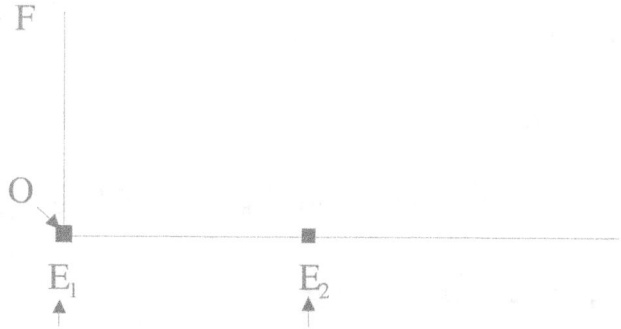

1st event at $x = t = 0$ in F   2nd event at a different place and different time in F

20

Let us introduce a second inertial frame, F', moving with speed V relative to F, in the +x

direction:

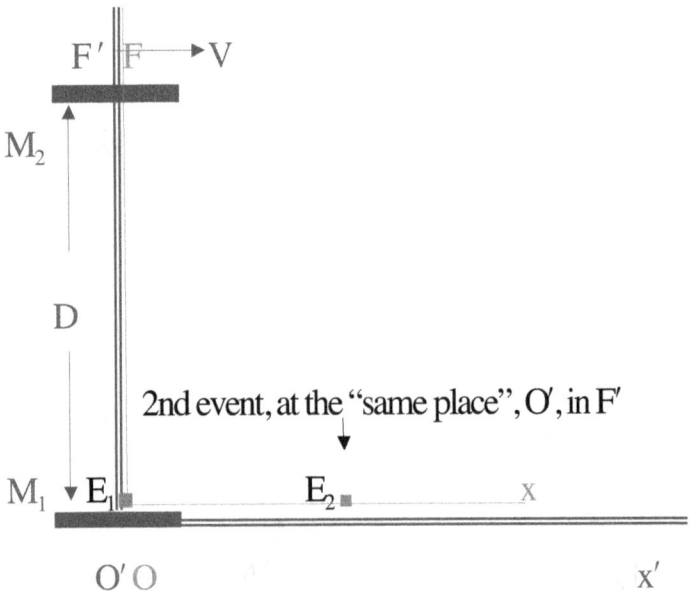

Let F ' be chosen in such a way that its origin, O', coincides with the first event E$_1$ (in both

space and time), and let its speed V be chosen so that O' coincides with the *location* of the

second event, E$_2$. We then have the situation in which *both events occur at the same place in*

F' (at the origin, O'). The interval between E$_1$ and E$_2$ in F ' is therefore a *proper interval*.

The mirrors, M$_1$ and M$_2$, are at rest in F ', with M$_1$ at the origin, O'. These mirrors move to the

right with speed V. Let a pulse of light be sent from the lower mirror when O and O' coincide

(at the instant that E$_1$ occurs). Furthermore, let the distance D between the mirrors be adjusted

so that the pulse returns to the lower mirror at the exact time and place of the event, E$_2$. This

sequence of events, as observed in F, is as shown:

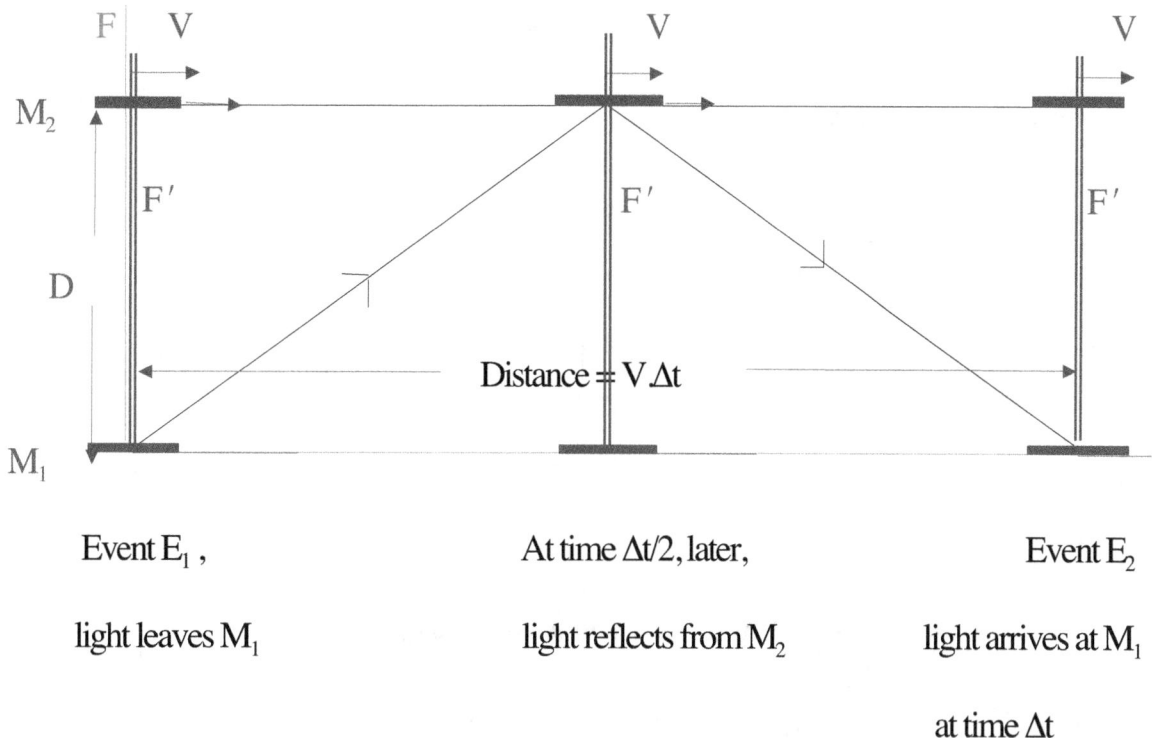

Event E_1,   At time $\Delta t/2$, later,   Event E_2

light leaves M_1 light reflects from M_2 light arrives at M_1

at time Δt

The sequence, observed in F′ is

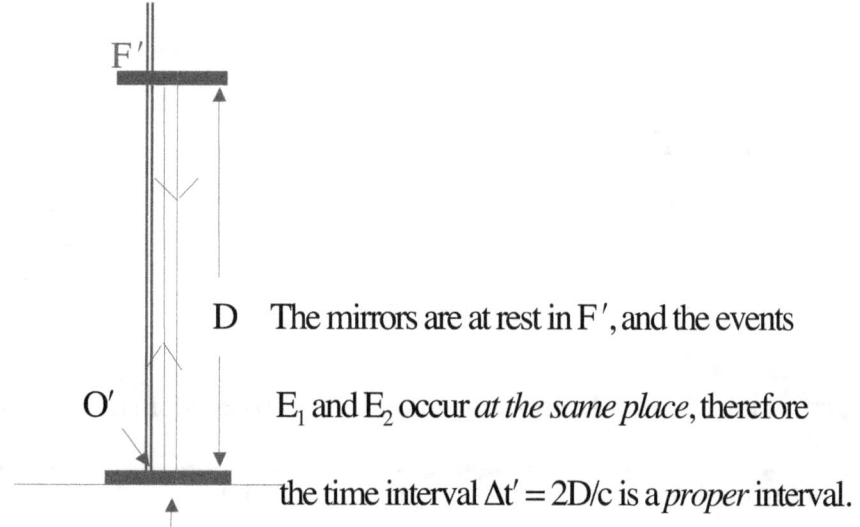

D   The mirrors are at rest in F′, and the events

O′ E_1 and E_2 occur *at the same place*, therefore

the time interval $\Delta t' = 2D/c$ is a *proper* interval.

E_1 and E_2, both take place at the origin, O′

The geometry of the sequence of events in the F frame is Pythagorean.

22

Recall, that in a right-angled triangle, we have

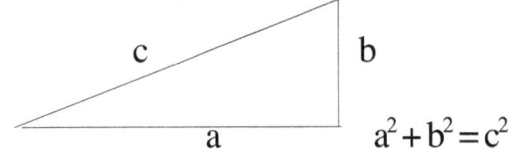

$$a^2 + b^2 = c^2.$$

In F, the relevant distances are:

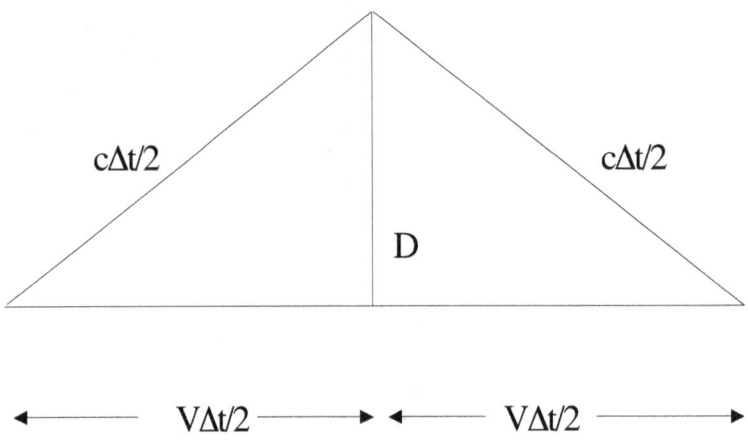

where Δt is the round-trip travel time. We therefore have

$$(V\Delta t/2)^2 + D^2 = (c\Delta t/2)^2,$$

or,

$$(\Delta t)^2[c^2 - V^2] = 4D^2,$$

so that, on taking the square root,

$$\Delta t = 2D/\sqrt{(c^2 - V^2)} = \{1/\sqrt{[1 - (V/c)^2]}\}(2D/c) = \gamma(2D/c),$$

where γ is the factor first introduced, on empirical grounds, by Fitzgerald. In this discussion, we

see that it emerges in a natural way from the two postulates of Einstein.

We can now compare the interval, Δt, between E_1 and E_2 in F with the interval, $\Delta t'$, between the same events as determined in F'. If we look at the sequence of events in F', in which the mirrors are at rest, we have

$$2D = c\Delta t',$$

and therefore

$$\Delta t' = 2D/c.$$

Substituting this value in the value for Δt in F, we obtain

$$\Delta t = \{1/\sqrt{[1 - (V/c)^2]}\}\Delta t'$$

or

$$\Delta t = \gamma\Delta t' , \gamma > 1,$$

which means that Δt (moving) $> \Delta t'$ (at rest).

A moving clock runs more slowly than an equivalent clock at rest.

Notice that at everyday speeds, in which V/c is typically less than 10^6, (and therefore $(V/c)^2$ is less than 10^{-12}, an unimaginably small number), Δt and $\Delta t'$ are essentially the same. Einstein's result then reduces to the classical result of Newton. However, in Modern Physics, involving microscopic particles that have measured speeds approaching that of light, values of $\gamma > 1000$ are often encountered. The equations of Newtonian Physics, and the philosophical basis of the equations, are then fundamentally wrong.

Although we have used an "optical clock" in the present discussion, the result applies to clocks in general, and, of course, to all inertial frames (they are equivalent).

A formal discussion of Einstein's Theory of Special Relativity is given in the Appendix.; it is intended for those with a flair for Mathematics.

4.4 Evidence for time dilation and length contraction

At the top of the Earth's atmosphere, typically 30,000m above sea-level, various gasses are found, including oxygen. When oxygen nuclei are bombarded with very high energy protons from the Sun, and from more distant objects, entities called muons are sometimes produced. These muons are found to have speeds very close to that of light (> 0.999c). Experiments show that the life-time of the muon, in its rest frame, is very short, a mere 2×10^{-6} seconds. After that brief existence, the muon transforms into other elementary particles. In Newtonian Physics, we would therefore expect the muon to travel a distance $d = V\Delta t$, where $V \approx c \approx 3 \times 10^{8}$ m/s, and $\Delta t = 2 \times 10^{-6}$ s, so that $d \approx 600$ m. We should therefore never expect to observe muons on the surface of the Earth, 30,000 m below. They are, however, frequently observed here on Earth, passing through us as part of the general cosmic background. Although the lifetime of the muon is 2μs in its rest frame, in the rest frame of the Earth, it is moving very rapidly, and therefore the interval between its creation and decay is no longer Δt but rather

$$\Delta t_E = \gamma \Delta t_\mu = 2 \times 10^{-6} \text{ seconds} / \sqrt{[1 - (V/c)^2]}$$

where V is the speed of the muon relative to Earth, and Δt_μ is the lifetime of the muon in its rest frame. Δt_E is its lifetime in the Earth's frame.

We see that if $\gamma \geq 50$, the muon will reach the Earth. (For then, $\Delta t_E \geq 10^{-4}$ s, and therefore $H_E \geq 30{,}000$m). A value of $\gamma = 50$ corresponds to a muon speed $V = 0.9995c$, and this is consistent with observations.

The detection of muons on the surface of the Earth is direct evidence for time dilation.

Alternatively, we may consider the observation of muons at the Earth's surface in terms of length contraction, as follows:

If we view the Earth from the rest frame of the muon, 30,000m above the Earth, it is moving toward the muon with very high speed, V corresponding to a value $\gamma \geq 50$. According to Einstein, the distance to the Earth is contracted by a factor of $\gamma \geq 50$, so that the muon-Earth distance from the perspective of the muon is $H_\mu = 30000/\gamma \leq 30000/50 \leq 600$m. The Earth therefore reaches the muon before it decays.

This is direct evidence for length contraction.

Schematically, we have:

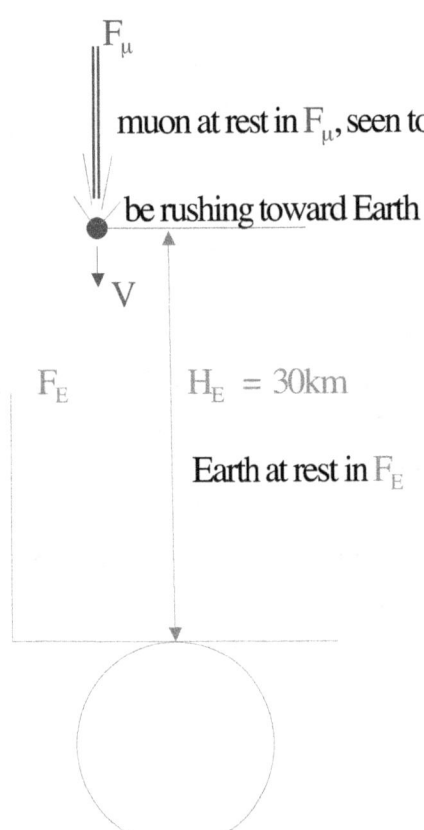

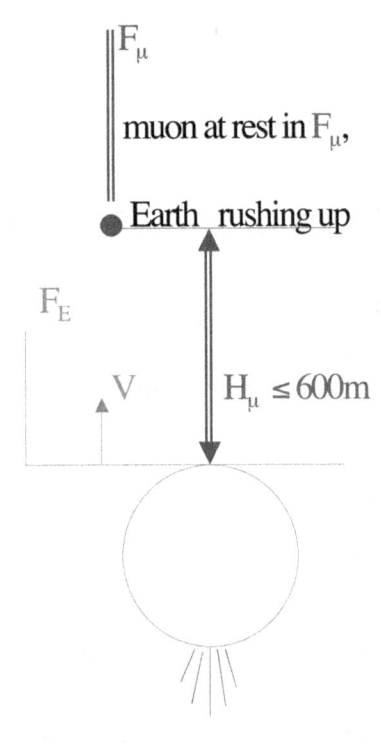

In F_E, fixed to the Earth, a muon

moving at high speed V toward

the Earth, is created at $H_E = 30$km.

In F_μ, fixed to the muon,

the Earth is moving

upwards at very high speed, V,

at a contracted distance

$H_\mu = \sqrt{1 - (V/c)^2}\; H_E.$

$= H_E / 50$ for $V = 0.9995c.$

4.5 Space travel

The distance between any two stars is so great that it is measured in "light-years", the distance light travels in one year. In more familiar units:

1 light-year $= 9.45 \times 10^{15}$ meters ≈ 6 trillion miles!

Alpha Centauri, our nearest star, is 4.3 light-years away; this means that, even if a spaceship could travel at a speed close to c, it would take more than 4.3 years to reach the star.

Imagine that it were technically possible to build such a near-light-speed craft; we ask how long the journey would take to a star, 80 light-years away, as measured by observers on the Earth. Let E_{LV} be the event "craft leaves Earth" and E_{ARR} the event "craft arrives at star". To observers in the rest frame of the Earth, it travels a distance 756×10^{15} meters at a speed of 0.995c, (say), in a time

$\Delta t_E = 756 \times 10^{15}$ (meters) $/ 0.995 \times 9.45 \times 10^{15}$ (meters per light-year)

$\quad = 80.4$ years.

Very few passengers would be alive when the craft reached the star. If the survivors sent a radio signal back to the Earth, saying that they had arrived, it would take another 80 years for the signal to reach the Earth. (Radio waves travel at the invariant speed of light).

Let us calculate the time of the trip from the perspective of the travelers. The spacecraft-fixed frame is labeled F_{SC}. In this frame, the two events, E_{LV} and E_{ARR} occur at the same place, namely the origin of F_{SC}. The trip-time, Δt_{SC}, according to the travelers is a "proper" time interval, and is less than the trip-time according to Earth-fixed observers by a factor $\sqrt{(1 - (V/c)^2)}$, where $V = 0.995c$. We then find

$$\Delta t_{SC} = \Delta t_E \sqrt{(1 - (0.995)^2)} = \Delta t_E \times \sqrt{(0.01)}$$

$$= 80.4 \times 0.1$$

$$= 8.04 \text{ years.}$$

To the travelers on board, the trip takes a little more than 8 years.

The passengers and crew of the spacecraft would have plenty of time to explore their new environment. Their return trip would take another 8.04 years, and therefore, on arrival back on the Earth, they would have aged 16.08 years, whereas their generation of Earth-bound persons would have long-since died. (The society would have aged 160 years).

This discussion assumes that the biological processes of the travelers take place according to time on the spacecraft clocks. This is certainly reasonable because heart-beats represent crude clocks, and metabolic rates of life processes are clock-like. According to Einstein, all "clocks" are affected by the motion of one inertial frame relative to another.

The twin paradox is not a paradox at all. Consider a twin, A, in the rest frame of the Earth, and let A observe the round-trip of his twin, B, to a distant star at near-c speed. A concludes that when B returns to Earth, they are no longer the same age; B is younger than A. According to the Principle of Relativity, B can be considered at rest, and the Earth, and twin A, travel away and return, later. In this case, B concludes that A has been on the trip, and therefore comes back younger. This cannot be! The paradox stems from the fact that the twin,

B, who leaves the Earth, must *accelerate* from the Earth, must *slow down* at the star, *turn around*, accelerate away from the star, and slow down on reaching Earth. In so doing, the B-twin has shifted out of inertial frames and into accelerating frames. Special Relativity does not hold throughout the entire journey of the twin in the spacecraft. The A-twin is always at rest in the inertial frame of the Earth. There is a *permanent asymmetry* in the space-time behavior of the twins.

The time in which a space traveler is in non-inertial frames can be made very short compared with the total travel time. The principles of Special Relativity are then valid, and the discussion given above, in which a spacecraft travels to a distant star, 80 light-years away, is essentially correct.

5. NEWTONIAN DYNAMICS

. . .for the whole burden of (natural) philosophy seems to consist of this

—from the phenomena of motions to investigate the forces of nature, and then from these forces to demonstrate the other phenomena.

NEWTON, the *PRINCIPIA*

Although our discussion of the geometry of motion has led to major advances in our understanding of measurements of space and time in different inertial systems, we have yet to come to the real crux of the matter, namely, a discussion of the effects of forces on the motion of two or more interacting particles. This key branch of Physics is called Dynamics. It was founded by Galileo and Newton and perfected by their followers, notably Lagrange and

Hamilton. The Newtonian concepts of mass, momentum and kinetic energy require fundamental revisions in the light of the Theory of Special Relativity. In spite of the conceptual difficulties inherent in the classical theory, its success in accounting for the dynamical behavior of systems, ranging from collisions of gas molecules to the motions of planets has been, and remains, spectacular.

5.1 The law of inertia

Galileo (1564-1642) was the first to develop a quantitative approach to the study of motion of everyday objects. In addition to this fundamental work, he constructed one of the first telescopes and used it to study our planetary system. His observation of the moons of Jupiter gave man his first glimpse of a miniature world system that confirmed the concepts put forward previously by Copernicus (1473-1543).

Galileo set out to answer the question: what property of motion is related to force? Is it the position of the moving object or its velocity or its rate of change of velocity, or what? The answer to this question can only be obtained from observations, this is a basic feature of Physics that sets it apart from Philosophy proper. Galileo observed that force Influences changes in velocity (accelerations) of an object and that, in the absence of external forces (e.g. friction), no force is needed to keep an object in motion that is traveling in a straight line with constant speed. This observationally based law is called the *Law of Inertia*. It is, perhaps, difficult for us to appreciate the impact of Galileo's new ideas concerning motion. The fact that an object resting on a horizontal surface remains at rest unless something we call force is applied to change its state of rest was, of course, well-known before Galileo's time. However, the fact that the object continues to move after the force ceases to be applied caused considerable conceptual difficulties to the early Philosophers. The observation that, in practice, an object comes to rest due to frictional forces and air resistance was recognized by Galileo to be a side effect and not germane to the fundamental question of motion. Aristotle, for example, believed that the true or natural state of motion is one of rest. It is instructive to consider Aristotle's conjecture from the viewpoint of the Principle of Relativity: is a natural state of rest consistent with this general Principle? First, we must consider what is meant by a natural state of rest; it means that in a particular frame of reference, the object in question is stationary. Now, according to the general Principle of Relativity, the laws of motion have the same form in all frames of reference that move with constant speed in straight lines with respect to each other. An observer in a reference frame moving with constant speed in a straight line with respect to the reference frame in which the object is at rest, would conclude that the natural state of motion of the object is one of constant speed in a straight line and not one of rest. All inertial observers, in an infinite number of frames of reference, would come to the same conclusion. We see, therefore, that Aristotle's conjecture is not consistent with this fundamental Principle.

5.2 Newton's laws of motion

During his early twenties, Newton postulated three laws of motion that form the basis of Classical Dynamics. He used them to solve a wide variety of problems, including the motion of the planets. They play a fundamental part in his famous Theory of Gravitation. The laws of motion were first published in the *Prlncipia* in l687; they are:

1. *In the absence of an applied force, an object will remain at rest or in its present state of constant speed in a straight line (Galileo's Law of Inertia)*

2. *In the presence of an applied force, an object will be accelerated in the direction of the applied force and the product of its mass by its acceleration is equal to the force.*

and,

3. *If a body A exerts a force of magnitude $|\mathbf{F}_{AB}|$ on a body B, then B exerts an equal force of magnitude $|\mathbf{F}_{BA}|$ on A. The forces act in opposite directions so that*

$$\mathbf{F}_{AB} = \mathbf{F}_{BA}.$$

The 3rd. Law applies to "contact interactions". For non-contact interactions, it is necessary to introduce the concept of a "field-of-force" that "carries the interaction".

We note that in the 2nd law, the acceleration lasts only while the applied force lasts. The applied force need not, however, be constant in time; the law is true at all instants during the motion. We can show this explicitly by writing:

$$\mathbf{F}(t) = m\mathbf{a}(t)$$

where the time-dependence of the force, and the resulting acceleration, is emphasized.

The "*mass*" appearing in Newton's 2nd law is the so-called *inertial mass*. It is that property of matter that resists changes in the state of motion of the matter. Later, in discussions of Gravitation, we shall meet another property of matter that also has the name "*mass*"; it is that property of matter that responds to the gravitational force due to the presence of other "masses"; this "*mass*" is the so-called "*gravitational mass*". The equivalence of inertial and gravitational mass was known to Newton. Einstein considered the equivalence to be of such fundamental importance that he used it as a starting point for his General Theory of Relativity, one of the greatest creations of the human mind.

In 1665 - 66, Sir Isaac Newton, the supreme analytical mind to emerge from England, deduced the basic law governing the interaction between two masses, M_1 and M_2. The force depends on the *product of the two masses*, and the *square of the distance between them*, thus

The gravitational force between the masses is given by:

$$F_{grav} \propto \frac{M_1 \times M_2}{R^2}$$

If the masses are initially 1 meter apart, and we increase the separation to 2 meters, the force decreases by $1/(2)^2 = 1/4$.

(It took Newton many years to prove that the distributed mass of a sphere can be treated as a "point" mass at its center. The problem involves a three-dimensional integral using his newly-invented Calculus).

5.3 General features of inverse square-law forces

In the early 1700's, Coulomb deduced the law of force that governs the interaction between two objects that possess the attribute of "electric charge". He found the following: the force between two charges Q_1 (at rest) and Q_2 depends on the *product of the two charges*, Q_1 and Q_2, and on the *square of the distance between them:*

Charge Q_1 Charge Q_2

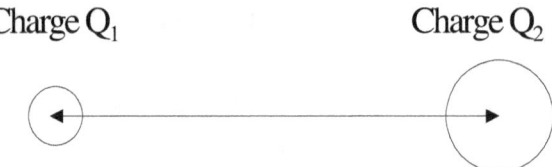

Distance between centers, R

$$F_{elect} \propto \frac{Q_1 \times Q_2}{R^2}$$

In the 19th-century, experiments showed that charges in motion, relative to an observer, generate an additional component of the force called the *Magnetic Force*. The complete force between moving charges is therefore known as the *Electromagnetic Force*.

We see that there is a remarkable similarity between the forms of the Gravitational and the Electromagnetic forces. They both depend on the symmetries $M_1 \times M_2$ and $Q_1 \times Q_2$, and they both vary as the inverse square of the distance between the objects. This latter feature is not by chance.

Let us introduce a model of these interactions in which we postulate that the force between one object and another is "carried", or mediated, by entities, generated by their sources; let them travel in straight lines between the objects. The mediators are capable of transferring momentum between the interacting objects. Consider the case in which a stationary charge Q is the source of mediators that travel, isotropically, from the charge at a uniform rate:

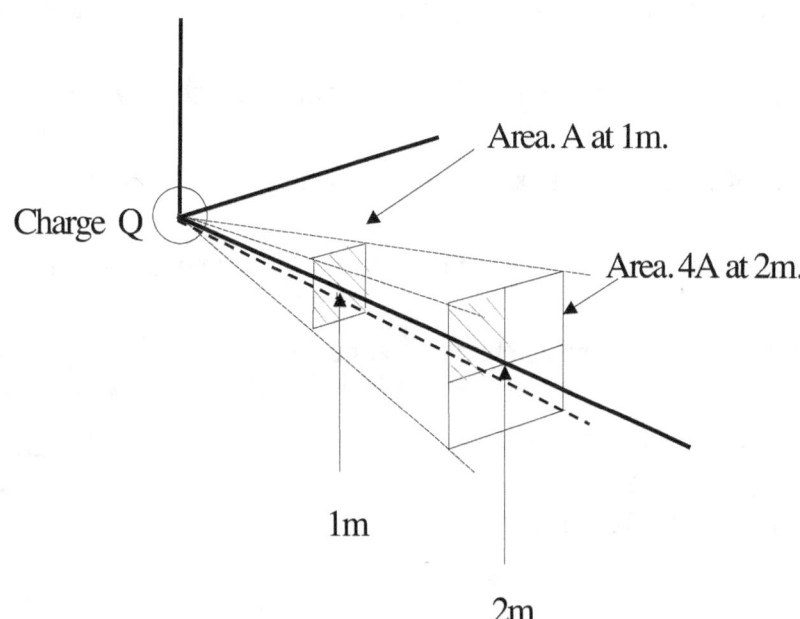

It is a property of the space in which we live that the shaded area, A, situated 1 meter from the charge Q, projects onto an area 4A at a distance 2 meters from Q. Therefore, the number of mediators passing through the area A at 1 meter from Q passes through an area 4A at a distance 2 meters from Q. If the force on a second charge Q′, 1 meter from Q, is due to the momentum per second transferred to the area A as a result of the mediators striking that area, then a charge Q′, 2 meters from the source, will experience a force that is 1/4 the force at 1 meter because the number of mediators per second passing through A now passes through an area four times as great. We see that the famous "inverse square law" is basically geometric in origin.

Not surprisingly, the real cases are more subtle than implied by this model. We must recognize a fundamental difference between the gravitational and the electromagnetic forces, namely:

the gravitational force is always attractive, whereas the electromagnetic force can be either attractive or repulsive. This difference comes about because there is only "one kind" of mass, whereas there are "two kinds" of charge, which we label positive and negative. (These terms were introduced by the versatile Ben Franklin). The interaction between like charges is repulsive and that between unlike charges is attractive.

We can develop our model of forces transmitted by the exchange of entities between objects that repel each other in the following way:

Consider two boxes situated on a sheet of ice. Let each box contain a person and a supply of basketballs. If the two occupants throw the balls at each other in such a way that they can be caught, then a stationary observer, watching the exchange, would see the two boxes moving apart. (This is a consequence of a law of motion that states that the linear momentum of a system is conserved in the absence of external forces):

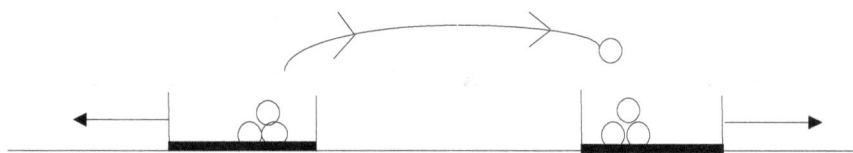

If the observer were so far away that he could see the boxes, but not the balls being exchanged, he would conclude that an unexplained *repulsive* force acted between the boxes.

The exchange model of an attractive force requires more imagination; we must invoke the exchange of *boomerangs* between the occupants of the boxes, as shown

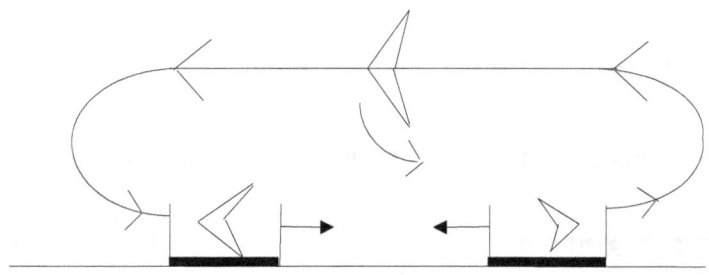

Boxes move together

These models are highly schematic. Nonetheless, they do indicate that models based on the exchange of entities that carry momentum, can be constructed. Contemporary theories of the Nuclear Force, and the Quark- Quark Force of Particle Physics, involve the exchange of exotic entities (mesons and gluons, respectively).

Newton deduced the inverse square law of gravitation by combining the results of painstaking observations of the motions of the planets (Brahe and Kepler), with an analysis of the elliptical motion of a (terrestrial) object, based on his laws of motion. This was the first time that the laws of motion, discovered locally, were applied on a universal scale.

6. EQUIVALENCE OF MASS AND ENERGY: $E = mc^2$

6.1 Relativistic mass

In Newtonian Physics, the inertial mass of an object is defined, operationally, by the second law:

$$m = F/a,$$

where a is the acceleration of the mass m, caused by the force F. For a given particle, the mass is constant; it has the same value in all inertial frames.

In Einsteinian Physics, the inertial mass, m, of an object depends on the speed of the frame in which it is measured. If its mass is m_0 in its rest frame then its mass m in an inertial frame moving at constant speed V is

$$m \text{ (the relativistic mass)} = \gamma m_0 \text{ (the rest mass)},$$

where γ is the same factor found in discussions of length contraction, and time dilation. The validity of this equation has been demonstrated in numerous modern experiments.

The structure of atoms has been understood since the early 1930's. An atom consists of a very small, positively charged nucleus, surrounded by electrons (negatively charged). The nucleus contains protons (positively charged) and neutrons (electrically neutral) bound together by the *nuclear force*. The diameter of a typical nucleus is less than 10^{-12} centimeters. The electrons orbit the nucleus at distances that can be one hundred thousand times greater than the size of the nucleus. The electrons are held in orbit by the *electromagnetic force*. The total positive charge of the nucleus is exactly balanced by the total negative charge of the planetary electrons, so that the atom is electrically neutral. An electron can be removed from an atom in different ways, including ionization in an electric field and photo-ionization with light. A free electron has the following accurately measured properties:

$$\text{mass of the electron} = 9.1083 \times 10^{-31} \text{ kilograms (kg)},$$

and

$$\text{electric charge of the electron} = 1.60206 \times 10^{-19} \text{ Coulomb. (C)}.$$

Here, "mass" refers to the mass measured in a frame of reference in which the electron is at rest; it is the "rest mass".

In 1932, a particle was observed with the same mass as the electron and with a charge equal in magnitude, but opposite in sign, to that of the electron. The particle was given the name "positron"; it is the "anti-particle" of the electron. The concept of anti-particles was introduced by Dirac using purely theoretical arguments, a few years before the experimental observation of the positron.

In the 1940's, it was found that an electron and a positron, when relatively at rest, may form a "positronium atom" that consists of a *bound state* of an electron and a positron, orbiting about their center-of-mass. Such an "atom" exists for a very short interval of time, namely 10^{-10} seconds. It then spontaneously decays into two gamma-rays (high energy electromagnetic radiation). The two gamma-rays are observed to travel back-to-back. (This observation is consistent with the law of conservation of linear momentum; the total momentum is zero before the decay (the particles are initially at rest), and therefore it must be zero after the decay). Each gamma-ray has a measured energy of 0.511 MeV (Million electron-volts).

In Modern Physics, it is the custom to use the electron-volt (eV) as the unit of energy. An electron-volt is the energy acquired by an electron when accelerated by a potential difference of one volt. Pictorially, we have:

(back-to-back)

− +   0.511 MeV 0.511 MeV

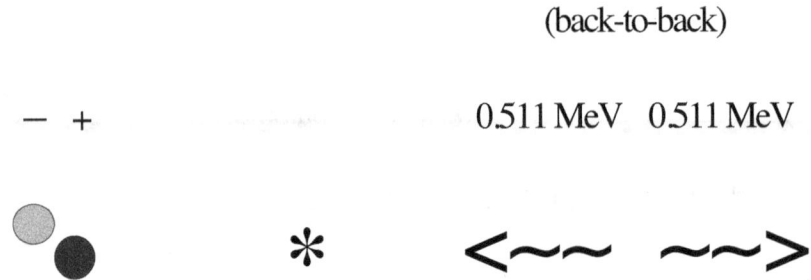

Matter − Anti-matter ⇒ Annihilation ⇒ Radiation

Electron-positron annihilation is a prime example of the conversion of matter into electromagnetic radiation. (This is the basic process in PET scanning [Positron Emission Tomography] in Nuclear Medicine).

From the measured electron and positron masses, and the measured energies of the two gamma-rays, we can obtain one of the most important results in our on-going quest for an understanding of the laws of Nature, and the associated workings of the physical universe. We wish to calculate the ratio

gamma-ray energy/electron mass.

The measured gamma-ray energy is:

$$E = 0.511 \times 10^6 \, (eV) \times 1.602 \times 10^{-19} \, (Joule / eV)$$

$$= 0.8186 \times 10^{-13} \, Joule.$$

(A note on "units": in the Physical Sciences, units of measured quantities are often given in the MKS system, in which *length*s are given in meters, *masses* are given in kilograms, and *time* is given in seconds. In this system, the unit of energy is the Joule (named after James Prescott Joule, a Manchester brewer and distinguished scientist of the 19th-century)).

Our required conversion factor is: 1 electron - volt = 1.602×10^{-19} Joule.

In the MKS system, the measured ratio gamma-ray energy/ electron mass is therefore

$$E / m = 0.8186 \times 10^{-13} \, Joule / 9.1083 \times 10^{-31} \, kilogram,$$

$$= 0.89874 \times 10^{17} \, (meters / second)^2,$$

a velocity, squared.

We can find the velocity by taking the square root of the value of E/m, thus:

$$\sqrt{(8.9874 \times 10^{16})} = 2.9974 \times 10^8 \text{ meters / second.}$$

This is a remarkable result; it is the exact value of the measured velocity of light, always written, c.

We therefore find that the ratio

$$E \text{ (gamma-ray)} / m \text{ (electron)} = c^2$$

or,

$$E = mc^2.$$

This is a particular version of Einstein's universal equation that represents the equivalence of energy and mass. It is important to note that Einstein derived this fundamental relation using purely theoretical arguments, before experiments were carried out to verify its validity. The heat that we receive from the Sun originates in the conversion of its central, highly compressed mass into radiant energy. A stretched spring has more mass than an unstretched spring, and a charged car battery has more mass than an uncharged battery! In both cases, the potential energy stored in the systems has an equivalent mass. We do not experience these effects because the mass changes are immeasurably small, due to the $1/c^2$ factor. However, in nuclear reactions that take place in nuclear reactors, or in nuclear bombs, the mass (energy) differences are enormous, and certainly have observable effects.

7. AN INTRODUCTION TO EINSTEINIAN GRAVITATION

7.1 The principle of equivalence

The term "mass" that appears in Newton's equation for the gravitational force between two interacting masses refers to

"gravitational mass"; Newton's law should indicate this property of matter

$$F_G = GM^G m^G / r^2,$$ where M^G and m^G are the *gravitational* masses of the interacting objects, separated by a distance r.

The term "mass" that appears in Newton's equation of motion, $F = ma$, refers to the "inertial mass"; Newton's equation of motion should indicate this property of matter:

$$F = m^I a,$$ where m^I is the *inertial* mass of the particle moving with an acceleration $a(r)$ in the gravitational field of the mass M^G.

Newton showed by experiment that the inertial mass of an object is equal to its gravitational mass, $m^I = m^G$ to an accuracy of 1 part in 10^3. Recent experiments have shown this equality to be true to an accuracy of 1 part in 10^{12}. Newton therefore took the equations

$$F = GM^G m^G / r^2 = m^I a$$

and used the condition $m^G = m^I$ to obtain

$$a = GM^G / r^2.$$

Galileo had previously shown that objects made from different materials fall with the same acceleration in the gravitational field at the surface of the Earth, a result that implies $m^G \propto m^I$. This is the Newtonian Principle of Equivalence.

Einstein used this Principle as a basis for a new Theory of Gravitation. He extended the axioms of Special Relativity, that apply to field-free frames, to frames of reference in "free fall". A freely falling frame must be in a state of *unpowered motion in a uniform gravitational field* . The field region must be sufficiently small for there to be no measurable gradient in the field throughout the region. The results of all experiments carried out in ideal freely falling frames are therefore fully consistent with Special Relativity. All freely falling observers measure the speed of light to be c, its constant free-space value. *It is not possible to carry out experiments in ideal freely-falling frames that permit a distinction to be made between the acceleration of local, freely-falling objects, and their motion in an equivalent external gravitational field.* As an immediate consequence of the extended Principle of Equivalence, Einstein showed that a beam of light would be deflected from its straight path in a close encounter with a sufficiently massive object. The observers would, themselves, be far removed from the gravitational field of the massive object causing the deflection.

Einstein's original calculation of the deflection of light from a distant star, grazing the Sun, as observed here on the Earth, included only those changes in *time intervals* that he had predicted would occur in the near field of the Sun. His result turned out to be in error by exactly a factor of two. He later obtained the "correct" value for the deflection by including in the calculation the changes in *spatial intervals* caused by the gravitational field.

7.2 Rates of clocks in a gravitational field

Let a rocket be moving with constant acceleration **a**, in a frame of reference, F, far removed

from the Earth's gravitational field, and let the rocket be instantaneously at rest in F at time t =

0. Suppose that two similar clocks, 1 and 2, are attached to the rocket with 1 at the rear end and

2 at the nose of the rocket. The clocks are separated by a distance ℓ. We can choose two light

sources, each with well-defined frequency, f_0, as suitable clocks. f_0 is the frequency when the

rocket is at rest in an inertial frame in free space.

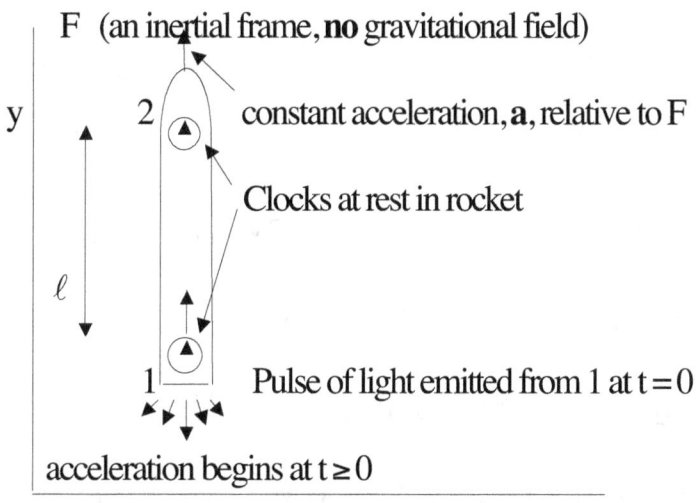

Let a pulse of light be emitted from the lower clock, 1, at time t = 0, when the rocket is

instantaneously at rest in F. This pulse reaches clock 2 after an interval of time t, (measured in

F) given by the standard equation for the distance traveled in time t:

$$ct = (\ell + (1/2)at^2),$$

where $(1/2)at^2$ is the extra distance that clock 2 moves in the interval t.

Therefore,

46

$$t = (\ell / c) + (a / 2c)t^2 \, ,$$

$$\approx (\ell / c) \text{ if } (at / 2) \ll c$$

At time t, clock 2 moves with velocity equal to $v = at \approx a\ell / c$, in F.

An observer at the position of clock 2 will conclude that the pulse of light coming from clock 1 had been emitted by a source moving downward with velocity v. The light is therefore "Doppler-shifted", the frequency is given by the standard expression for the Doppler shift at low speeds ($v \ll c$):

$$f' \approx f_0[1 - (v / c)]$$

$$= f_0[1 - (a\ell / c^2].$$

The frequency f' is therefore less than the frequency f_0. The light from clock 1 (below) is "red-shifted". Conversely, light from the upper clock traveling down to the lower clock is measured to have a higher frequency than the local clock 1; it is "blue-shifted".

The *principle of equivalence* states that the above situation, in a closed system, cannot be distinguished by physical measurements, from that in which the rocket is at rest in a uniform gravitational field. The field must produce an acceleration of magnitude |a|, on all masses placed in it.

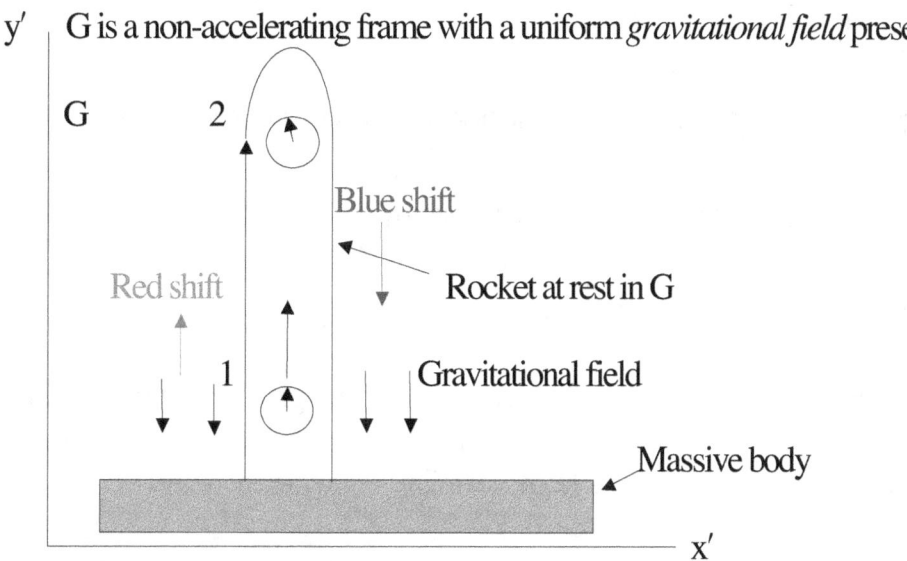

The light from the lower clock, reaching the upper clock will have a frequency lower than the local clock, 2, by $f_0 g\ell / c^2$, (replacing |a| by |g|), where $g \approx 10$ m / s^2, the acceleration due to gravity near the Earth. The light sources are at rest in G, and no oscillations of the pulses of light are lost during transmission; we therefore conclude that, in a uniform gravitational field, the *actual* frequencies of the stationary clocks differ by $f_0 g\ell / c^2$. Now, $g\ell$ is the difference in the "gravitational potential" between the two clocks. It is the convention to say that the upper clock, 2, is at the higher potential in G. (Work must be done to lift the mass of clock 1 to clock 2 against the field).

Consider the case in which a light source of frequency f_S (corresponding to clock 1) is situated on the surface of a star, and consider a similar light source on the Earth with a frequency f_E (corresponding to clock 2). Generalizing the above discussion to the case when the two clocks are in *varying* gravitational fields, such that the difference in their potentials is $\Delta\phi$, we find

$$f_E = f_S(1 + \Delta\phi / c^2)$$

$(g\ell = \Delta\phi$, is the difference in gravitational potential of the clocks in

a constant field, g, when separated by ℓ).

For a star that is much more massive than the Earth, $\Delta\phi$ is positive, therefore, $f_E > f_S$, or in terms of wavelengths, λ_E and λ_S, $\lambda_S > \lambda_E$. This means that the light coming from the distant star is red-shifted compared with the light from a similar light source, at rest on the surface of the Earth.

As another example, radioactive atoms with a well-defined "half-life" should decay faster near clock 2 (the upper clock) than near clock 1. At the higher altitude (higher potential), all physical processes go faster, and the frequency of light from above is higher than the frequency of light from an identical clock below. Einstein's prediction was verified in a series of accurate experiments, carried out in the late 1950's, using radioactive sources that were placed at different heights near the surface of the Earth.

7.3 Gravity and photons: red shifts, blue shifts and black holes

Throughout the 19th-century, the study of optical phenomena, such as the diffraction of light by an object, demonstrated conclusively that light (electromagnetic in origin) behaves as a wave. In 1900, Max Planck, analyzed the results of experimental studies of the characteristic spectrum of electromagnetic radiation emerging from a hole in a heated cavity (so-called "black-body radiation"). He found that current theory, that involved *continuous* frequencies in the spectrum, could not explain the results. He did find that the main features of all black-body spectra could be explained by making the radical assumption that the radiation consists of discrete pulses of energy E proportional to the frequency, f. By fitting the data, he determined the constant of proportionality, now called Planck's constant; it is always written h. The present value is:

$$h = 6.626 \times 10^{-34} \text{ Joule - second in MKS units.}$$

Planck's great discovery was the beginning of Quantum Physics.

In 1905, Einstein was the first to apply Planck's new idea to another branch of Physics, namely, the Photoelectric Effect. Again, current theories could not explain the results. Einstein argued that discrete pulses of electromagnetic energy behave like *localized particles*, carrying energy E = hf and momentum p = E/c. These particles interact with tiny electrons in the surface of metals, and eject electrons in a Newtonian-like way. He wrote

$$E_{PH} = hf_{PH} \text{ and } E_{PH} = p_{PH}c$$

The rest mass of the photon is zero. (Its energy is all kinetic).

If, under certain circumstances, photons behave like particles, we are led to ask: are photons affected by gravity? We have

$$E_{PH} = m_{PH}{}^{I} c^2 = hf_{PH},$$

or

$$m_{PH}{}^{I} = E_{PH}/c^2 = hf_{PH}/c^2.$$

By the Principle of Equivalence, inertial mass is equivalent to gravitational mass, therefore Einstein proposed that a beam of light (photons) should be deflected in a gravitational field, just as if it were a beam of particles. (It is worth noting that Newton considered light to consist of particles; he did not discuss the properties of his particles. In the early 1800's, Soldner actually calculated the deflection of a beam of "light-particles" in the presence of a massive object! Einstein was not aware of this earlier work).

Let us consider a photon of initial frequency f_S, emitted by a massive star of mass M_S, and radius R. The gravitational potential energy, V, of a mass m at the surface of the star, is given by a standard result of Newton's Theory of Gravitation; it is

$$V(\text{surface}) = -GM_S m / R.$$

It is inversely proportional to the radius of the star. The negative sign signifies that the gravitational interaction between M_S and m is always *attractive*.

Following Einstein, we can write the potential energy of the photon of "mass" hf_{PH}/c^2 at the surface as

$$V(\text{surface}) = -(GM_S/R)(hf_{PH}/c^2).$$

The total energy of the photon, E_{TOTAL} is the sum of its kinetic and potential energy:

$$E_{TOTAL} = hf_{PH}^{STAR} + (-)GM_S hf_{PH}^{STAR}/Rc^2,$$

$$= hf_{PH}^{STAR}(1 - GM_S/Rc^2).$$

At very large distances from the star, at the Earth, for example, the photon is essentially beyond the gravitational pull of the star. Its total energy remains unchanged (conservation of energy). At the surface of the Earth the photon has an energy that is entirely electromagnetic (since its potential energy in the "weak" field of the Earth is negligible compared with that in the gravitational field of the star), therefore

$$hf_{PH}^{EARTH} = hf_{PH}^{STAR}(1 - GM_S/Rc^2)$$

so that

$$f_{PH}^{EARTH}/f_{PH}^{STAR} = 1 - GM_S/Rc^2,$$

and

$$\Delta f/f \equiv (f_{PH}^{STAR} - f_{PH}^{EARTH})/f_{PH}^{STAR} = GM_S/Rc^2 .$$

We see that the photon on reaching the Earth has less total energy than it had on leaving the star. It therefore has a lower frequency at the Earth. If the photon is in the optical region, it is shifted towards the red-end of the spectrum. This is the *gravitational red-shift*. (It is quite different from the red-shift associated with Special Relativity)

Schematically, we have:

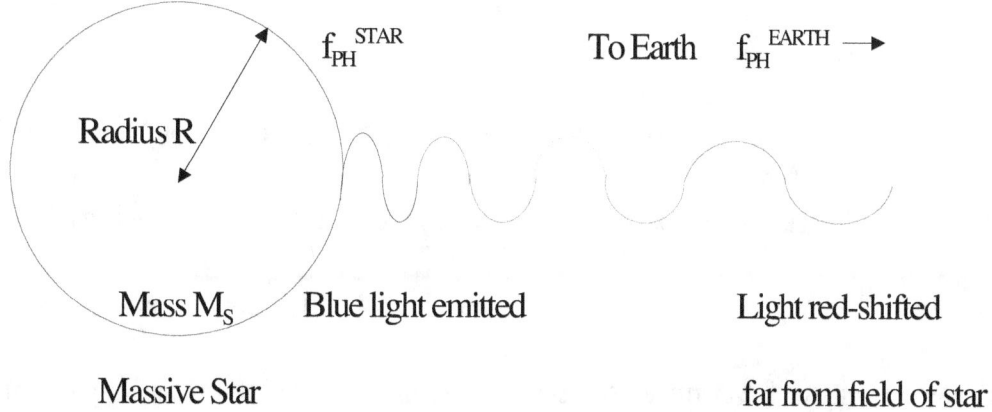

Radius R

f_{PH}^{STAR} To Earth $f_{PH}^{EARTH} \longrightarrow$

Mass M_S Blue light emitted   Light red-shifted

Massive Star   far from field of star

In 1784, a remarkable paper was published in the *Philosophical Transactions of the Royal Society of London*, written by the Rev. J. Michell. It contained the following discussion:

To escape to an infinite distance from the surface of a star of mass M and radius R, an object of mass m must have an initial velocity v_0 given by the energy condition:

initial kinetic energy of mass $\geq$ potential energy at surface of star,

or

$$(1/2)mv_0^2 \geq GMm / R \text{ (A Newtonian expression).}$$

This means that

$$v_0 \geq \sqrt{(2GM/R)}.$$

Escape is possible only when the initial velocity is greater than $(2GM/R)^{1/2}$.

On the Earth, $v_0 \geq 25{,}000$ miles/hour.

For a star of given mass M, the escape velocity *increases* as its radius decreases. Michell considered the case in which the escape velocity v_0 reaches a value c, the speed of light. In this limit, the radius becomes

$$R_{LIMIT} = 2GM / c^2$$

He argued that light would not be able to escape from a compact star of mass M with a radius less than R_{LIMIT}; the star would become invisible. In modern terminology, it is a *black hole*.

Using the language of Einstein, we would say that the curvature of space-time in the immediate vicinity of the compact star is so severe that the time taken for light to emerge from the star becomes infinite. The radius $2GM / c^2$ is known as the *Schwarzschild radius*; he was the first to obtain a particular solution of the Einstein equations of General Relativity. The analysis given by Michell, centuries ago, was necessarily limited by the theoretical knowledge of his day. For example, his use of a non-relativistic expression for the kinetic energy ($mv^2/2$) is now known to require modification when dealing with objects that move at speeds close to c. Nonetheless, he obtained an answer that turned out to be essentially correct. His use of a theoretical argument based on the conservation of energy was not a standard procedure in Physics until much later.

A star that is 1.4 times more massive than our Sun, has a Schwarzschild radius of only 2km and a density of 10^{20} kg/m^3. This is far greater than the density of an atomic nucleus. For more compact stars ($R_{LIMIT} < 1.4\ M_{SUN}$), the gravitational self-attraction leads inevitably to its collapse to a "point".

Studies of the X-ray source Cygnus X-1 indicate that it is a member of a binary system, the other member being a massive "blue supergiant". There is evidence for the flow of matter from the massive optical star to the X-ray source, with an *accretion disc* around the center of the X-ray source. The X-rays could not be coming from the blue supergiant because it is too cold. Models of this system, coupled with on-going observations, are consistent with the conjecture that a black hole is at the center of Cygnus X-1. Several other good candidates for black holes have been observed in recent studies of binary systems. The detection of X-rays from distant objects has become possible only with the advent of satellite-borne equipment.

I have discussed some of the great contributions made by Einstein to our understanding of the fundamental processes that govern the workings of our world, and the universe, beyond. He was a true genius.

APPENDIX

The following material presents the main ideas of Einstein's Special Relativity in a mathematical form. It is written for those with a flair for Mathematics.

A 1. Some useful mathematics: transformations and matrices

Let a point P[x, y] in a Cartesian frame be *rotated* about the origin through an angle of 90°; let the new position be labeled P′[x′, y′]

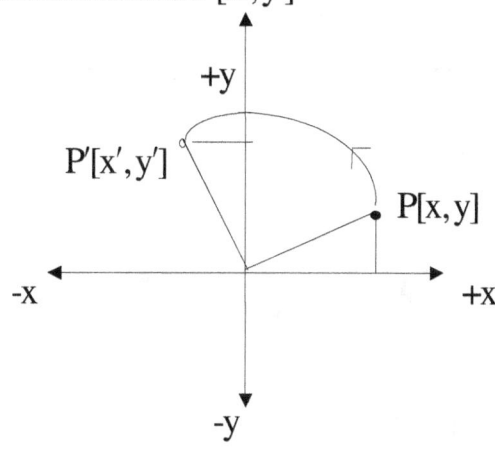

We see that the new coordinates are related to the old coordinates as follows:

$$x' \text{ (new)} = \qquad -y \text{ (old)}$$

and

$$y' \text{ (new)} = +x \text{ (old)}$$

where we have written the x's and y's in different *columns* for reasons that will become clear, later.

Consider a *stretching* of the material of the plane such that all x-values are doubled and all y-values are tripled:

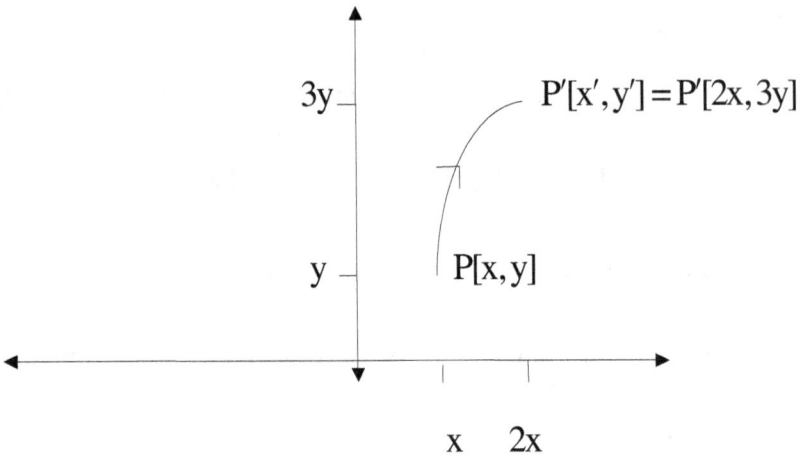

The old coordinates are related to the new coordinates by the equations

$$y' = 3y$$

and

$$x' = 2x$$

Consider a more complicated transformation in which the new values are combinations of the old values, for example, let

$$x' = 1x + 3y$$

and

$$y' = 3x + 1y$$

We can see what this transformation does by putting in a few definite values for the coordinates:

$$[0,0] \rightarrow [0,0]$$

$$[1,0] \rightarrow [1.1 + 3.0, 3.1 + 1.0] = [1,3]$$

$$[2,0] \rightarrow [1.2 + 3.0, 3.2 + 1.0] = [2,6]$$

$$[0,1] \rightarrow [1.0 + 3.1, 3.0 + 1.1] = [3,1]$$

$$[0,2] \rightarrow [1.0 + 3.2, 3.0 + 1.2] = [6,2]$$

$$[1,1] \rightarrow [1.1 + 3.1, 3.1 + 1.1] = [4,4\}$$

$$[1,2] \rightarrow [1.1 + 3.2, 3.1 + 1.2] = [7,5]$$

$$[2,2] \rightarrow [1.2 + 3.2, 3.2 + 1.2] = [8,8]$$

$$[2,1] \rightarrow [1.2 + 3.1, 3.2 + 1.1] = [5,7]$$

and so on.

Some of these changes are shown below

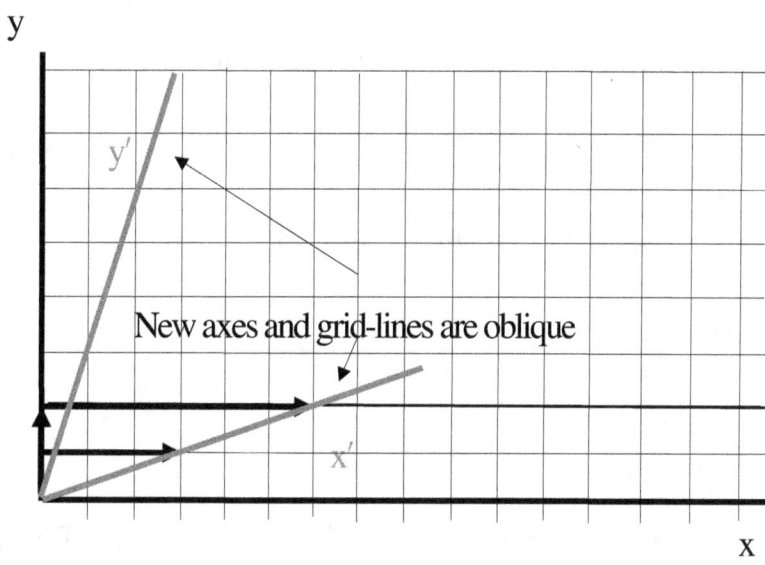

This is a particular example of the more general transformation

$$x' = ax + by$$

and

$$y' = cx + dy$$

where a, b, c, and d are real numbers.

In the above examples, we see that each transformation is characterized by the values of the coefficients, a, b, c, and d:

For the *rotation* through 90°:

$a = 0, b = -1, c = 1$, and $d = 0$;

for the 2x3 *stretch*:

$a = 2, b = 0, c = 0$, and $d = 3$;

and for the more general transformation:

$a = 1, b = 3, c = 3$, and $d = 1$.

In the 1840's, Cayley recognized the key role of the coefficients in characterizing the transformation of a coordinate pair [x, y] into the pair [x′, y′]. He therefore "separated them out", writing the pair of equations in column-form, thus:

$$\begin{pmatrix} x' \\ y' \end{pmatrix} = \begin{pmatrix} a & b \\ c & d \end{pmatrix} \begin{pmatrix} x \\ y \end{pmatrix}$$

This is a *single* equation that represents the original *two* equations. We can write it in the symbolic form:

$$\mathbf{P'} = \mathbf{MP} ,$$

which means that the point **P** with coordinates x, y (written as a column) is changed into the point **P′** with coordinates x′, y′ by the operation of the **2 x 2 matrix operator M**.

The matrix **M** is

$$\mathbf{M} = \begin{pmatrix} a & b \\ c & d \end{pmatrix}.$$

The algebraic rule for carrying out the "matrix multiplication" is obtained directly by noting that the single symbolic equation is the equivalent of the two original equations. We must therefore have

$$x' = a \text{ times } x + b \text{ times } y$$

and

$$y' = c \text{ times } x + d \text{ times } y.$$

We *multiply rows of the matrix* by *columns of the coordinates*, in the correct order.

2 x 2 matrix operators will be seen to play a crucial role in Einstein's Special Theory of Relativity.

A 2. Galilean-Newtonian relativity revisited

The idea of matrix operators provides us with a useful way of looking at the equations of classical relativity, discussed previously. Recall the two basic equations:

$$t' = t$$

and

$$x' = x - Vt \ .$$

where, the event **E**[t, x] in the F-frame has been transformed into the event **E**′[t′, x′] in the F′-frame. We can write these two equations as a *single* matrix operator equation as follows

$$\begin{pmatrix} t' \\ x' \end{pmatrix} = \begin{pmatrix} 1 & 0 \\ -V & 1 \end{pmatrix} \begin{pmatrix} t \\ x \end{pmatrix},$$

or, symbolically

$$\mathbf{E}' = \mathbf{GE},$$

where

$$\mathbf{G} = \begin{pmatrix} 1 & 0 \\ -V & 1 \end{pmatrix}, \text{the matrix of the Galilean transformation.}$$

If we transform $\mathbf{E} \to \mathbf{E}'$ under the operation $\mathbf{G}$, we can undo the transformation by carrying out the **inverse operation**, written $\mathbf{G}^{-1}$, that transforms $\mathbf{E}' \to \mathbf{E}$, by *reversing* the direction of the relative velocity:

$$t = t'$$

and

$$x = x' + Vt'$$

or, written as a matrix equation:

$$\begin{pmatrix} t \\ x \end{pmatrix} = \begin{pmatrix} 1 & 0 \\ +V & 1 \end{pmatrix} \begin{pmatrix} t' \\ x' \end{pmatrix}$$

where

$$\mathbf{G}^{-1} = \begin{pmatrix} 1 & 0 \\ +V & 1 \end{pmatrix}$$

is the **inverse** operator of the Galilean transformation. Because $\mathbf{G}^{-1}$ undoes

the effect of $\mathbf{G}$, we have

$$\mathbf{G}^{-1}\mathbf{G} = \text{``do nothing''} = \mathbf{I}\text{, the }\textit{identity}\text{ operator,}$$

where

$$\mathbf{I} = \begin{pmatrix} 1 & 0 \\ 0 & 1 \end{pmatrix}.$$

We can illustrate the space-time path of a point moving with respect to the F- and F'-

frames *on the same graph,* as follows

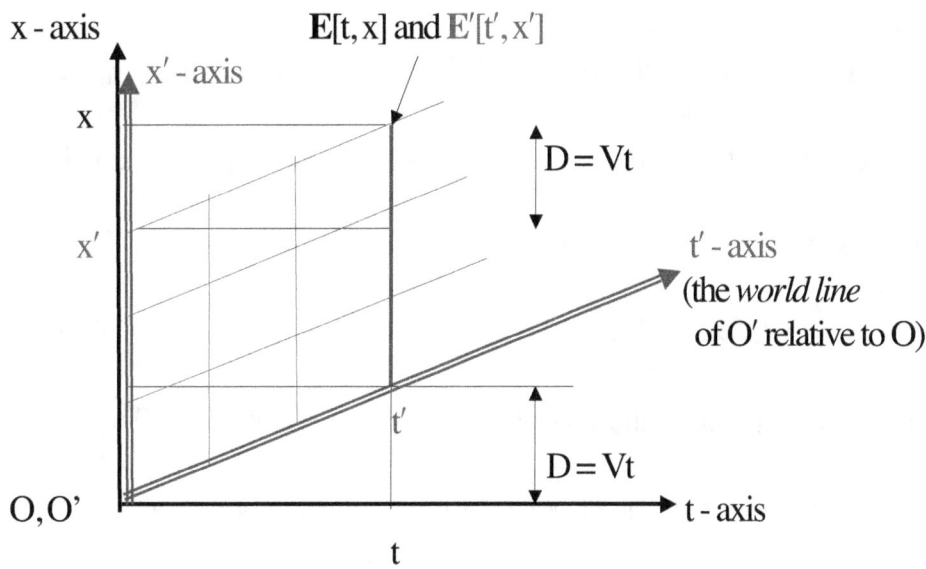

The origins of F and F' are chosen to be coincident at $t = t' = 0$. O' moves to the right

with constant speed V, and therefore travels a distance $D = Vt$ in time t. The t' - axis is the

world line of O' in the F-frame. Every point in this space-time geometry obeys the relation

$x' = x - Vt$; the F'-frame is therefore represented by a *semi-oblique coordinate system.* The

characteristic feature of Galilean-Newtonian space-time is the coincidence of the x-x'- axes.
Note that the time intervals, t, t' in F and F' are numerically the same (Newton's "absolute
time"), and therefore a new time scale must be chosen for the oblique axis, because the *lengths*
along the time-axis, corresponding to the times t, t' of the event **E**, **E'** are different.

A 3. Is the geometry of space-time Pythagorean?

Pythagoras' Theorem is of primary importance in the geometry of space. The theorem
is a consequence of the invariance properties of lengths and angles under the operations of
translations and rotations. We are therefore led to ask the question – do invariants of space-
time geometry exist under the operation of the Galilean transformation and, if they do, what are
they? We can address this question by making a simple, direct calculation, as follows. The
basic equations that relate space-time measurements in two inertial frames moving with
relative speed V are

$$x' = x - Vt \text{ and } t' = t.$$

We are interested in quantities of the form $x^2 + t^2$ and $x'^2 + t'^2$. These forms are inconsistent,
however, because the "dimensions" of the terms are not the same; x, x' have dimensions of
"length" and t, t' have dimensions of "time". This inconsistency can be dealt with by
introducing two quantities k, k' that have dimensions "length/time" (speed), so that the
equations become

$$x' = x - Vt \text{ (all lengths) and } k't' = kt \text{ (all lengths)}.$$

(Note that kt is the *distance* traveled in a time t at a constant speed k). We now find

$$x'^2 = (x - Vt)^2 = x^2 - 2xVt + V^2t^2, \text{ and } k'^2t'^2 = k^2t^2,$$

so that

$$x'^2 + k'^2t'^2 = x^2 - 2xVt + V^2t^2 + k^2t^2.$$

$$\neq x^2 + k^2t^2 \text{ unless } V = 0 \text{ (no motion!)}.$$

Relative events in an semi-oblique space-time geometry therefore transform under the Galilean operator in a non-Pythagorean way.

A 4. Einstein's space-time symmetry: the Lorentz transformation

We have seen that the classical equations relating the events $\mathbf{E}$ and $\mathbf{E}'$ are

$\mathbf{E}' = \mathbf{GE}$, and the inverse $\mathbf{E} = \mathbf{G}^{-1}\mathbf{E}'$ where

$$\mathbf{G} = \begin{pmatrix} 1 & 0 \\ -V & 1 \end{pmatrix} \quad \text{and} \quad \mathbf{G}^{-1} = \begin{pmatrix} 1 & 0 \\ V & 1 \end{pmatrix}.$$

These equations are connected by the substitution $V \leftrightarrow -V$; this is an algebraic statement of the Newtonian *Principle of Relativity*. Einstein incorporated this principle in his theory (his *first* postulate), broadening its scope to include *all* physical phenomena, and not simply the motion of mechanical objects. He also retained the *linearity* of the classical equations in the absence of any evidence to the contrary. (Equispaced intervals of time and distance in one inertial frame remain equispaced in any other inertial frame). He therefore *symmetrized* the space-time equations (by putting *spac*e and *time* on equal footings) as follows:

64

$$
\begin{pmatrix} t' \\ \\ x' \end{pmatrix} = \begin{pmatrix} 1 & -V \\ \\ -V & 1 \end{pmatrix} \begin{pmatrix} t \\ \\ x \end{pmatrix} .
$$

−V replaces 0, to *symmetrize* the matrix

Note, however, the inconsistency in the dimensions of the time-equation that has now been introduced:

$$t' = t - Vx.$$

The term Vx has dimensions of $[L]^2/[T]$, and not $[T]$. This can be corrected by introducing the *invariant* speed of light, c (Einstein's *second* postulate, consistent with the result of the Michelson-Morley experiment):

$$ct' = ct - Vx/c \quad (c' = c, \text{ in all inertial frames})$$

so that all terms now have dimensions of *length*. (ct is the distance that light travels in a time t)

Einstein went further, and introduced a dimensionless quantity γ instead of the scaling factor of unity that appears in the Galilean equations of space-time. (What is the number "1" doing in a theory of space-time?). This factor must be consistent with all observations. The equations then become

$$ct' = \gamma ct - \beta\gamma x$$

$$x' = -\beta\gamma ct + \gamma x , \text{ where } \beta = V/c.$$

These can be written

$$\mathbf{E'} = \mathbf{LE},$$

where

$$\mathbf{L} = \begin{pmatrix} \gamma & -\beta\gamma \\ -\beta\gamma & \gamma \end{pmatrix},$$

and

$$\mathbf{E} = [ct, x].$$

$\mathbf{L}$ is the operator of *the Lorentz transformation.* (First obtained by Lorentz, it is the transformation that leaves Maxwell's equations of electromagnetism unchanged in form between inertial frames).

The inverse equation is $\mathbf{E} = \mathbf{L}^{-1}\mathbf{E}'$, where

$$\mathbf{L}^{-1} = \begin{pmatrix} \gamma & \beta\gamma \\ \beta\gamma & \gamma \end{pmatrix}.$$

This is the *inverse* Lorentz transformation, obtained from $\mathbf{L}$ by changing $\beta \rightarrow -\beta$ ($V \rightarrow -V$); it has the effect of undoing the transformation $\mathbf{L}$. We can therefore write

$$\mathbf{L}\mathbf{L}^{-1} = \mathbf{I}, \text{ the identity.}$$

Carrying out the matrix multiplication, and equating elements gives

$$\gamma^2 - \beta^2\gamma^2 = 1$$

therefore,

$$\gamma = 1/\sqrt{(1-\beta^2)} \text{ (taking the positive root).}$$

As $V \rightarrow 0, \beta \rightarrow 0$ and therefore $\gamma \rightarrow 1$; this represents the *classical* limit in which the Galilean transformation is, for all practical purposes, valid. In particular, time intervals have the same measured values in all Galilean frames of reference, and *acceleration* is the single Galilean

invariant.

A 5. The invariant interval

Previously, it was shown that the space-time of Galileo and Newton is not Pythagorean

under **G**. We now ask the question: is Einsteinian space-time Pythagorean under **L** ? Direct

calculation leads to

$$(ct)^2 + x^2 = \gamma^2(1 + \beta^2)(ct')^2 + 4\beta\gamma^2 x' ct'$$

$$+\gamma^2(1 + \beta^2)x'^2$$

$$\neq (ct')^2 + x'^2 \text{ if } \beta > 0.$$

Note, however, that the ***difference of squares*** is an ***invariant***:

$$(ct)^2 - x^2 = (ct')^2 - x'^2$$

because

$$\gamma^2(1 - \beta^2) = 1.$$

Space-time is said to be pseudo-Euclidean. *The "difference of squares" is the characteristic*

feature of Nature's space-time. The "minus" sign makes no sense when we try and relate it to

our everyday experience of geometry. The importance of Einstein's "free invention of the

human mind" is clearly evident in this discussion.

The geometry of the Lorentz transformation, **L**, between two inertial frames involves

oblique coordinates, as follows:

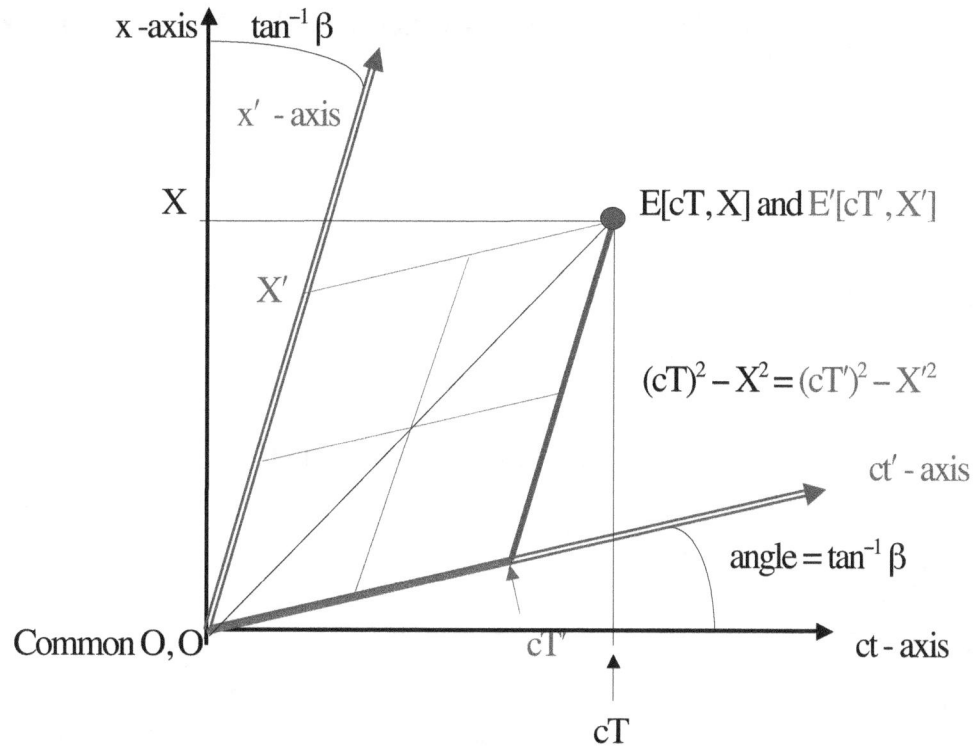

The symmetry of space-time means that the ct - axis and the x - axis fold through *equal* angles.

Note that when the relative velocity of the frames is equal to the speed of light, c, the folding

angle is 45^0, and the space-time axes coalesce.

A 6. The relativity of simultaneity: the significance of oblique axes

Consider two sources of light, 1 and 2, and a point M midway between them. Let E_1

denote the event "flash of light leaves 1", and E_2 denote the event "flash of light leaves 2". The

events E_1 and E_2 are *simultaneous* if the flashes of light from 1 and 2 reach M at the same time.

The oblique coordinate system that relates events in one inertial frame to the same events in a

second (moving) inertial frame shows, in a most direct way, that two events observed to be

coincident in one inertial frame are not observed to be coincident in a second inertial frame (moving with a constant relative velocity, $\mathbf{V}$, in standard geometry). Two events $E_1[ct_1, x_1]$ and

$E_2[ct_2, x_2]$, are observed in a frame, F. Let them be coincident in F, so that $t_1 = t_2 = t$, (say). The

two events are shown in the following diagram:

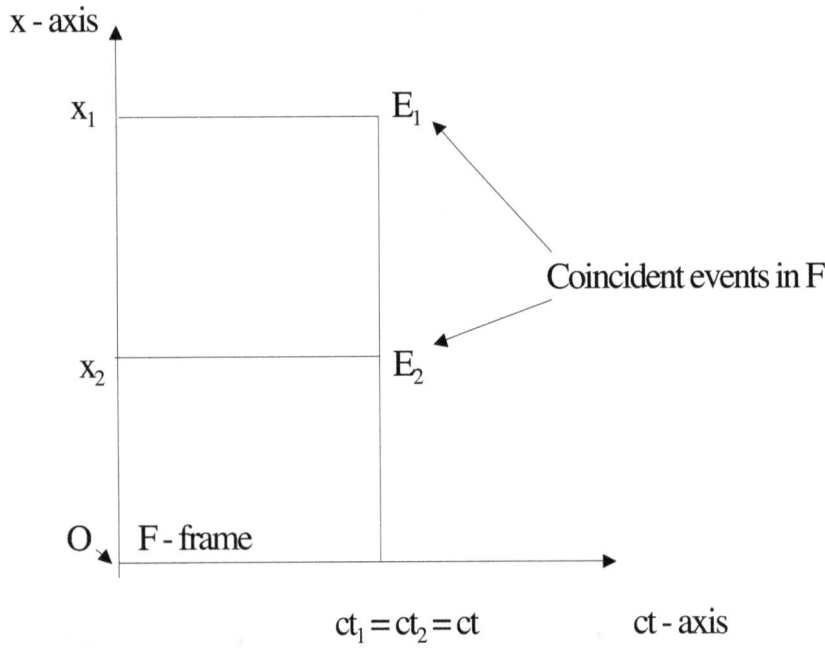

Consider the same two events as measured in another inertial frame, F', moving at constant

velocity $\mathbf{V}$ along the common positive x - x' axis. In F', the two events are labeled $E_1'[ct_1', x_1']$

and $E_2'[ct_2', x_2']$. Because F' is moving at constant velocity $+\mathbf{V}$ relative to F, the space-time

axes of F' are folded inwards through angles $\tan^{-1}(V/c)$ relative to the F axes, as shown. The

events E_1' and E_2' can be displayed in the F' - frame:

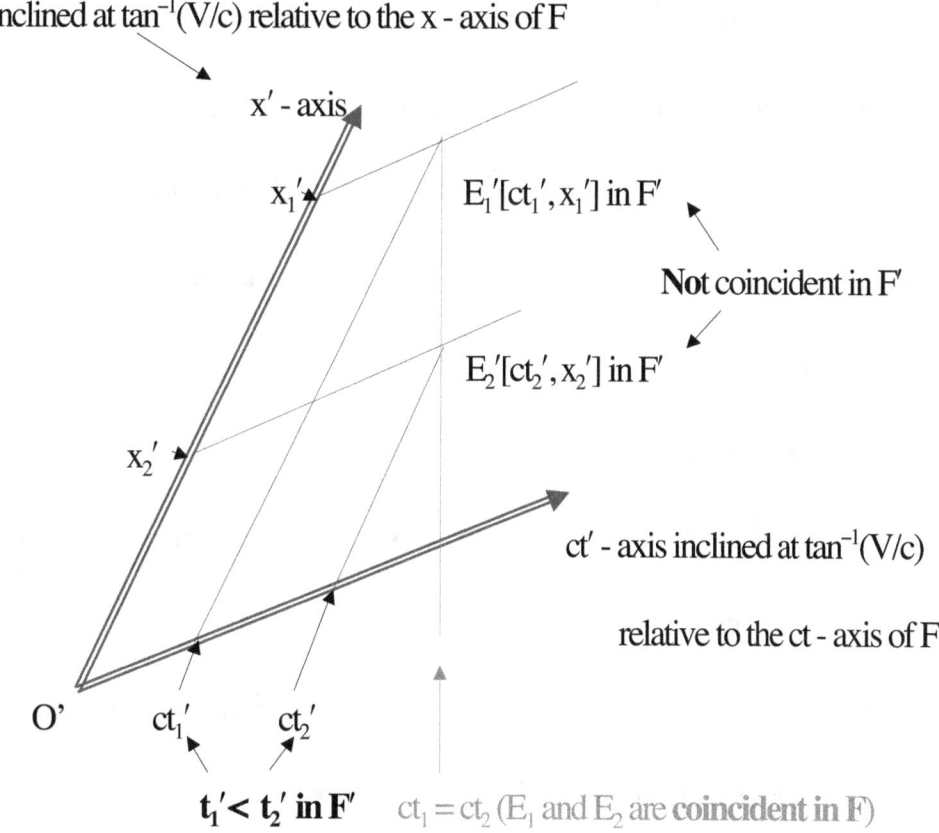

inclined at $\tan^{-1}(V/c)$ relative to the x - axis of F

x′ - axis

x_1'

$E_1'[ct_1', x_1']$ in F′

Not coincident in F′

$E_2'[ct_2', x_2']$ in F′

x_2'

ct′ - axis inclined at $\tan^{-1}(V/c)$

relative to the ct - axis of F

O′ ct_1' ct_2'

$\mathbf{t_1' < t_2'}$ **in F′** $ct_1 = ct_2$ (E_1 and E_2 are **coincident in F**)

We therefore see that, for all values of the relative velocity $V > 0$, the events E_1' and E_2' as measured in F′ are not coincident; E_1' occurs before E_2'.

(If the sign of the relative velocity is reversed, the axes fold outwards through equal angles).

A 7. Length contraction: the Lorentz transformation in action

The measurement of the length of a rod involves comparing the two ends of the rod with marks on a standard ruler, or some equivalent device. If the object to be measured, and the ruler, are at rest in our frame of reference then it does not matter when the two end-positions are determined - the "length" is clearly-defined. If, however, the rod is in motion, the meaning of its *length* must be reconsidered. *The positions of the ends of the rod relative to the standard ruler must be "measured at the same time" in its frame of reference.*

Consider a rigid rod at rest on the x'-axis of an inertial reference frame F'. Because it is at rest, it does not matter when its end-points x_1' and x_2' are measured to give the rest-, or *proper*-length of the rod, $L_0' = x_2' - x_1'$.

Consider the same rod observed in an inertial reference frame F that is moving with constant velocity $-\mathbf{V}$ with its x-axis parallel to the x'-axis. We wish to determine the length of the moving rod; We require the length $L = x_2 - x_1$ according to the observers in F. This means that the observers in F must measure x_1 and x_2 *at the same time* in their reference frame. The events in the two reference frames F, and F' are related by the spatial part of the Lorentz transformation:

$$x' = -\beta\gamma ct + \gamma x$$

and therefore

$$x_2' - x_1' = -\beta\gamma c(t_2 - t_1) + \gamma(x_2 - x_1).$$

where

$$\beta = V/c \text{ and } \gamma = 1/\sqrt{(1 - \beta^2)}.$$

Since we require the length $(x_2 - x_1)$ in F to be measured *at the same time* in F, we must have $t_2 - t_1 = 0$, and therefore

$$L_0' = x_2' - x_1' = \gamma(x_2 - x_1),$$

or

$$L_0'(\text{at rest}) = \gamma L \text{ (moving)}.$$

The length of a moving rod, L, is therefore less than the length of the same rod measured at rest, L_0, because $\gamma > 1$.

A 8. Time dilation: a formal approach

Consider a *single clock* at rest at the origin of an inertial frame F´, and a *set of synchronized clocks* at x_0, x_1, x_2, ... on the x-axis of another inertial frame F. Let F´ move at constant velocity +**V** relative to F, along the common x -, x´- axis. Let the clocks at x_0, and x_0' be synchronized to read t_0 and t_0' at the instant that they coincide in space. A *proper time interval* is defined to be the time between two events measured in an inertial frame in which they occur *at the same place*. The time part of the Lorentz transformation can be used to relate an interval of time measured on the single clock in the F´ frame, and the same interval of time measured on the set of synchronized clocks at rest in the F frame. We have

$$ct = \gamma ct' + \beta\gamma x'$$

or

$$c(t_2 - t_1) = \gamma c(t_2' - t_1') + \beta\gamma(x_2' - x_1').$$

There is no separation between a single clock and itself, therefore $x_2' - x_1' = 0$, so that

$$c(t_2 - t_1)(\text{moving}) = \gamma c(t_2' - t_1')(\text{at rest}) ,$$

or

$$c\Delta t \,(\text{moving}) = \gamma c\Delta t' \,(\text{at rest}).$$

Therefore, because $\gamma > 1$, a moving clock runs more slowly than a clock at rest.

A 9. Relativistic mass, momentum, and energy

The scalar product of a vector $\mathbf{A}$ with components $[a_1, a_2]$ and a vector $\mathbf{B}$ with components $[b_1, b_2]$ is

$$\mathbf{A} \cdot \mathbf{B} = a_1 b_1 + a_2 b_2.$$

In geometry, $\mathbf{A} \cdot \mathbf{B}$ is an *invariant* under rotations and translations of the coordinate system.

In *space-time*, Nature prescribes the differences-of-squares as the invariant under the Lorentz transformation that relates measurements in one inertial frame to measurements in another. For two events, $E_1[ct, x]$ and $E_2[ct, -x]$, the scalar product is

$$\mathbf{E}_1 \cdot \mathbf{E}_2 = [ct, x].[ct, -x]$$

$$= (ct)^2 - x^2 = \text{invariant in a space-time geometry,}$$

where we have chosen the direction of E_2 to be opposite to that of E_1, thereby providing the necessary negative sign in the invariant.

In terms of finite differences of time and distance, we obtain

$$(c\Delta t)^2 - (\Delta x)^2 \equiv (c\Delta \tau)^2 = \text{invariant,}$$

where $\Delta\tau$ is the proper time interval. It is related to Δt by the equation

$$\Delta t = \gamma \Delta \tau.$$

In Newtonian Mechanics, the quantity *momentum*, the product of the mass of an object and its velocity, plays a key role. In Einsteinian Mechanics, velocity, mass, momentum and kinetic energy are redefined. These basic changes are a direct consequence of the replacement of Newton's absolute time interval, Δt_N, by the Einstein's velocity-dependent interval

$$\Delta t_E = \gamma \Delta \tau.$$

The Newtonian momentum $\mathbf{p}_N = m_N \mathbf{v}_N = m_N \Delta \mathbf{x}/\Delta t_N$ is replaced by the Einsteinian momentum

$$\mathbf{p}_E^+ \equiv m_0 \mathbf{v}_E = m_0 \Delta[ct, x]/\Delta \tau$$

$$= m_0[c\Delta t/\Delta \tau, \Delta x/\Delta \tau]$$

$$= m_0[\gamma c, (\Delta x/\Delta t)(\Delta t/\Delta \tau]$$

$$= m_0[\gamma c, \gamma v_N].$$

We now introduce the vector in which the direction of the x-component is reversed, giving

$$\mathbf{p}_E^- = m_0[\gamma c, -\gamma v_N].$$

Forming the scalar product, we obtain

$$\mathbf{p}_E^+ \cdot \mathbf{p}_E^- = m_0^2(\gamma^2 c^2 - \gamma^2 v_N^2)$$

$$= m_0^2 c^2,$$

because $\quad \mathbf{v}_E^+ \cdot \mathbf{v}_E^- = c^2.$

Multiplying throughout by c^2, and rearranging, we find

$$m_0^2 c^4 = \gamma^2 m_0^2 c^4 - \gamma^2 m_0^2 c^2 v_N^2 .$$

We see that γ is a number and therefore γ multiplied by the rest mass m_0 is a *mass*; let us therefore denote it by m:

$$m = \gamma m_0, \text{ the relativistic mass.}$$

74

We can then write

$$(m_0c^2)^2 = (mc^2)^2 - (cp_E)^2.$$

The quantity mc^2 has dimensions of *energy*; let us therefore denote it by the symbol E, so that

$$E = mc^2, \text{Einstein's fundamental equation.}$$

The equivalence of mass and energy is seen to appear in a natural way in our search for the invariants of Nature.

The term involving m_0 is the *rest energy*, E_0,

$$E_0 = m_0c^2.$$

We therefore obtain

$$E_0^2 = E^2 - (p_Ec)^2 = E'^2 - (p_E'c)^2, \text{in any other inertial frame.}$$

It is the *basic invariant* of relativistic particle dynamics.

This invariant includes those particles with zero rest mass. For a photon of total energy E_{PH} and momentum p_{PH}, we have

$$0 = E_{PH}^2 - (p_{PH}c)^2,$$

and therefore

$$E_{PH} = p_{PH}c.$$

No violations of Einstein's Theory of Special Relativity have been found in any tests of the theory that have been carried to this day.

Bibliography

The following books are written in a style that requires little or no Mathematics:

Calder, N., *Einstein's Universe*, The Viking Press, New York (1979).

Davies, P. C. W., *Space and Time in the Modern Universe*,

Cambridge University Press, Cambridge (1977).

The following books are mathematical in style; they are listed in increasing level of mathematical sophistication:

Casper, Barry M., and Noer, Richard J., *Revolutions in Physics*,

W. W. Norton & Company Inc., New York (1972).

Born, M., *The Special Theory of Relativity*, Dover, New York (1962).

French, A. P., *Special Relativity*, W. W. Norton & Company, Inc.

New York (1968).

Rosser, W. G. V., *Introduction to Special Relativity*, Butterworth & Co. Ltd.

London (1967).

Feynman, R. P., Leighton, R. B., and Sands, M., *The Feynman Lectures on*

Physics, Addison-Wesley Publishing Company, Reading, MA (1964).

Rindler, W., *Introduction to Special Relativity*, Oxford University Press,

Oxford, 2nd ed. (1991).